Health and safety for engineers

Edited by Martin J. Barnard

Published by Thomas Telford Publishing, Thomas Telford Limited, 1 Heron Quay, London E14 4JD.
URL: http://www.t-telford.co.uk

Distributors for Thomas Telford books are
USA: ASCE Press, 1801 Alexander Bell Drive, Reston, VA 20191-4400
Japan: Maruzen Co. Ltd, Book Department, 310 Nihonbashi 2-chome, Chuo-ku, Tokyo 103
Australia: DA Books and Journals, 648 Whitehorse Road, Mitcham 3132, Victoria

First published 1998

Also available from Thomas Telford Books:
Health and safety law for the construction industry. ISBN 0 7277 2602 1
Construction safety handbook, 2nd edition. ISBN 0 7277 2519 X
Total project management of construction safety, health and environment, 2nd edition. ISBN 0 7277 1923 8

A catalogue record for this book is available from the British Library

ISBN: 0 7277 2602 1

Typeset by MHL Typesetting Limited, Coventry
Printed and bound in Great Britain by Bookcraft (Bath) Ltd

Preface

Most professionals in construction would tell you that there are three essential ingredients to a successful project: quality, time and cost. I passionately believe that there are in fact five, with safety and environment added to the list. Although not part of this book, we should not ignore the impact of the work we do on the environment, particularly from the eye of the lay-person. On the safety front I have seen numerous examples of projects where the desire to do the work safely has contributed to a successful project. We should not be surprised by this because the principles of good safety management are the same as good project management, that is the right level of emphasis on planning and preparation makes the execution straightforward and successful.

For many in the construction industry, health and safety has always been someone else's problem. Even very senior managers have declared a non-interest, often on the basis that that is what the Safety Officer is for. This dated concept is still sadly prevalent in many organizations. A growing number of enlightened ones are realizing that health and safety has to be an integral part of their business. It must feature from the outset of any project with a clear understanding of what the end result should be and a vision as to how that will be achieved. This book seeks to encourage the young engineer to recognize that the fundamental objective should be:

> *the right person in the right place at the right time doing the right job using the right equipment.*

Good health and safety management is something to which many organizations aspire, just as they do in the commercial sense. It is also something which the law requires. The clearest lesson is that the starting point is the genuine and thoughtful commitment of all concerned. I commend the principles in this book to build a safe and successful future for you. By doing so both the spirit and the letter of safety law can be met.

M.J. Barnard

Contents

Preface iii

1. Legal requirements. S.E. FINK 1
2. Professional ethics and morals. D.I. BLOCKLEY 14
3. The CDM Regulations: keeping an air of realism! M.J. BARNARD 26
4. Design for safety. M.A. WILLIAMS 40
5. Risk assessment. J. GREEN 56
6. Personal safety. E.M. BENNETT 79
7. Small capital and maintenance works. G.S.J. FULWELL 93
8. Managing hazardous substances in construction. N.E.G. MARTENS 114
9. Subcontract management. M.H. EVANS 134
10. International safety considerations. P.E. BROWN 149
11. The Health and Safety Commission and Executive: dealing with an inspector. S. NATTRASS 159
12. Twenty-one years on: construction and trade union representatives. T.J. MELLISH 173
13. Growing a safety culture. J. ANDERSON 185

1. Legal requirements

Susan E. Fink, *Masons Solicitors, London*

Bases of health and safety law

Statutory law

Health and safety statutes came into being well over two hundred years ago. Those early statutes typically were prescriptive and inflexible, leaving no scope for individual differences.

Despite those laws, accident rates continued to rise. That fact led the British government in the 1970s to commission the first comprehensive review of health and safety law in the UK, the Robens Report. That report suggested that health and safety law should move away from detailed and prescriptive rules to principles of good practice, which encouraged companies to set their own safety goals and to develop their own safety procedures. This recommendation in turn led to the creation of the Health and Safety at Work etc. Act 1974 (HSWA), which, we will see, imposes a wide range of duties on employers and others to ensure safety in the work place.

The Health and Safety at Work etc. Act 1974

The primary focus for all health and safety legislation in the UK is the HSWA and the subsidiary legislation passed thereunder. While there are other health and safety Acts still in force which could affect safety on a construction site (e.g. the Factories Act 1961 and the Offices, Shops and Railway Premises Act 1963), they are slowly being phased out and replaced by subsidiary legislation.

Employees' safety

Part 2 of the HSWA imposes a number of broad duties on employers to ensure the safety of their employees. Specifically that Part obliges employers, so far as is reasonably practicable, to:

- ensure the health and safety and welfare of their employees
- ensure the provision and maintenance of safe plant and systems of work for employees
- provide safe systems with regard to the use, storage and transport of articles and substances for employees

- ☐ provide such information, instruction, training and supervision as is necessary to ensure the health and safety at work of employees
- ☐ maintain a safe place of work and to provide and maintain safe access to and egress from that place of work for employees
- ☐ provide and maintain a working environment for employees that is safe without risk to health.

As we will see, these duties serve as the basis for most of the health and safety legislation that is now in force in the UK.

Safety of non-employees

In addition to the duties employers have to their employees, they also have duties to ensure the health and safety of persons *not* in their employ under Part 3 of the HSWA. Part 3(1) imposes a duty on employers to ensure, so far as is reasonably practicable, the health and safety of persons who are not his employees but who may be affected by the 'conduct of his undertaking'. This Part has particular relevance for the construction industry, in view of the industry's reliance on subcontracted labour and the likelihood of interaction with the general public.

The scope of this Part has been considered by the House of Lords recently in *R* v. *Associated Octel Company Ltd* [1996]. That case involved Octel, a chemical manufacturer which had shut its chemical plant down for its annual service and had employed a firm of specialist contractors to perform the maintenance and repair work. During one cleaning operation, a light bulb being used by one of the contractor's employees burst, igniting the cleaning fluid he was using and badly burning him. Both the contractor and Octel were prosecuted, with Octel convicted of failing to secure the safety of persons not in its employ under Part 3(1).

Octel appealed its conviction first to the Court of Appeal, then to the House of Lords. On 14 November 1996 the House of Lords rejected Octel's appeal and upheld the conviction. Octel had argued that it had no duty to ensure the health and safety of the subcontractor's employees under Part 3(1) because the activities of that employee were not within the conduct of its undertaking. In his majority opinion Lord Hoffman rejected that view, holding instead that there was no doubt that Octel was conducting its undertaking when it instructed the specialist contractor because:

> Octel's undertaking was running a chemical plant at Ellesmere Port. Anything which constituted running the plant was part of the conduct of its undertaking ... [As such] it is part of the conduct of the undertaking, not merely to clean the factory, but to have the factory cleaned by contractors.

Having decided that it applied, Lord Hoffman went on to decide that Part 3 obliges an employer to take all

> reasonably practicable steps to avoid risk to the contractors' servants which arise, not merely from the physical state of the premises ... but also from the inadequacy of the arrangements which the employer makes with the contractor for how they do the work.

To summarize: Part 3(1) obliges employers to take precautions, so far as is reasonably practicable, for the safety of persons not in his employ but who are affected by his undertaking. The Octel decision makes it clear that simply by instructing a subcontractor to do work associated with his business, an employer is 'conducting his undertaking' for purposes of Part 3, and as such has duties to ensure the safety of that contractor and his employees, both from the conditions on the premises and from any inadequacies in his safety management systems.

Statutory instruments, ACOP and guidance

The HSWA provides little in the way of detail or standards for employers to follow. Therefore, the Health and Safety Commission (HSC) has been empowered to develop subsidiary legislation to supplement and give depth to the duties outlined in the Act.

As a result, there is now a vast range of statutory instruments, regulations and orders in effect—far too many to discuss here. Having said that, there are a few statutes which require special mention in this chapter because of their particular relevance for the construction industry. They include:

- The Management of Health and Safety at Work Regulations 1992, as amended. These Regulations impose a number of duties on employers to conduct risk assessments of their operations and to develop procedures to limit risks identified; to co-operate and co-ordinate their safety efforts with other employers; to appoint persons who are competent to deal with safety on the site; to provide information and training to their employees; and to communicate their safety procedures to persons coming on to the site. Compliance with these Regulations should form a fundamental part of all employers' safety procedures
- The Provision and Use of Work Equipment Regulations 1992 and the Personal Protective Equipment at Work Regulations 1992, as amended, both impose duties on employers and others to assess the risks associated with the use of equipment at work to ensure that it is appropriate for the job and for the persons who use it; and to ensure that the equipment remains safe to use at all times.
- The Construction (Health, Safety and Welfare) Regulations 1996 sets out precise standards for employers and others who work on construction sites to meet. Those standards cover issues such as:
  - falling objects and fall prevention

- stability of structures
- demolition or dismantling
- explosives
- excavations
- vehicles and traffic routes
- emergency procedures, and
- welfare, training and information.

☐ The Manual Handling Operations Regulations 1992 oblige employers to assess the risk of injury associated with manual handling, and to take all reasonably practicable steps to limit those risks.

☐ The Construction (Head Protection) Regulations 1989 impose a duty on all persons working on a construction site to wear suitable head protection when at risk from falling materials.

This is far from a complete list of the relevant statutes; however, it does give the reader an idea as to the range of health and safety duties that exist.

In addition to statutes, the HSC has developed a number of Approved Codes of Practice (ACOP) and/or Guidance Notes to provide interpretative assistance and practical guidance for those trying to understand and comply with health and safety legislation. While a breach of an ACOP or a Guidance Note is not an offence, such a breach can be used as evidence of an offence. As such, compliance with such Codes and/or Guidance is strongly recommended.

'So far as is reasonably practicable'

It is import to note that, virtually without exception, the duties imposed by the HSWA and the subsidiary legislation passed thereunder are limited by the phrase 'so far as is reasonably practicable'. That phrase has the effect of permitting any person subject to a health and safety duty to conduct a cost/benefit analysis in respect of his health and safety systems. In other words, it allows the duty holder to calculate whether the benefits afforded (i.e. a reduction in the risk of injury) by the addition of health and safety procedures are outweighed by the costs (delays, inconvenience, additional expense, etc.) of such additional procedures. If so, those precautions need not, in theory, be taken.

The Construction (Design and Management) Regulations

It is fair to say that the Construction (Design and Management) Regulations 1994 (CDM Regulations) represent one of the most fundamental developments in construction safety law since the 1960s. In view of this, it is essential to consider these Regulations in some detail, beginning with a brief outline of their historical background.

The Temporary or Mobile Construction Sites Directive

The CDM Regulations implement the Temporary or Mobile Construction Sites Directive, which was passed by the European Council in 1992. That Directive was adopted in response to research which indicated that poor management of a construction project, and in particular poor training, communication and planning, were directly related to levels of risk on a construction site.

As a result of these findings, the Directive and the CDM Regulations have developed a safety management structure for construction sites which recognizes that:

- construction clients are in a very good position to exercise control over the levels of safety on a construction site
- designers have control over the safety of their designs and should be held accountable therefore, and
- contractors should be able to plan for safety from tender stage, through all stages of construction.

The following is a brief summary of the CDM Regulations. For more details, please refer to the ACOP and/or the Guidance Notes for the Regulations prepared by the Health and Safety Executive (HSE) in conjunction with the Construction Industry Advisory Council.

The CDM Regulations apply to virtually all 'construction work', which is broadly defined to include most building and engineering works, including new construction, repairs, maintenance, refurbishment or demolition, but excluding mineral extraction and exploration.

The Health and Safety Plan

The Health and Safety Plan is intended to make plain to contractors the health and safety risks associated with a construction project. The Plan is developed in two stages: the pre-tender plan and the final plan.

The pre-tender plan is to be organized by the Planning Supervisor (see below) as soon as possible after the project is conceived and should form part of the tender documentation. The plan should set out all of the significant safety risks associated with the proposed project to enable contractors to develop appropriate health and safety procedures and systems for the project. Once the Principal Contractor (see below) has been appointed, it becomes his responsibility to develop the plan to its final form. The ACOP and the Guidance Notes to the Regulations provide details on what the plan should contain at both the pre-tender and final stages.

The plan must be updated and amended whenever appropriate, and may be prepared on a phase-by-phase basis if that is how the construction work is to proceed. The ACOP suggests that any amendments made by the Principal Contractor to the plan that were developed by the Planning

Supervisor must be confirmed with the Planning Supervisor before being implemented.

The Health and Safety File

A Health and Safety File must be prepared for every *structure* that comprises a construction project. The File should contain information about a structure's design, construction and intended use so that any person planning construction in the future in respect of that structure will understand the hazards that may be associated therewith. Again, the ACOP and the Guidance Notes to the CDM Regulations make recommendations as to the contents of the File.

The Client

The Client is any person for whom construction work is being carried out, whether by external or in-house labour. This definition does not include a householder who contracts to have work done on his own home, however.

The Client must appoint a Planning Supervisor and a Principal Contractor to the project. This appointment should be made as soon as possible after the Client has satisfied himself of their 'competence' and that they have 'allocated adequate resources' to enable them to perform their functions.

The Client is expected to provide information about the state and condition of the construction site to the Planning Supervisor, as soon as practicable before work commences. That should include any information that would assist the Planning Supervisor in complying with his duties under the Regulations.

Also, the Client must ensure, so far as reasonably practicable, that an appropriate Health and Safety Plan has been developed by the Principal Contractor before permitting construction work to begin, but at the same time he must give the Principal Contractor sufficient time to develop the Plan before expecting construction to begin.

The Client must take reasonable steps to ensure that the information contained in the Health and Safety File is available for inspection by any person (e.g. subcontractors or the emergency services) who may need it to comply with their statutory duties. Ultimately, the Client must deliver the File (or a copy of the File) to any person acquiring an interest in the premises, such as purchaser or a tenant.

The Designer

The Designer is any person who prepares a design for construction, or who employs a person to prepare such designs. This broad definition would include engineers and architects, as well as quantity surveyors and project managers involved with design.

Before a Designer can begin a design for a Client, he must take steps to notify the Client of the latter's duties under the CDM Regulations. In this way, it is hoped that the Client will undertake his duties as quickly and effectively as possible, thereby making it easier for the Designer to do his job.

The most essential part of the Designer's duties under the Regulations is to ensure, so far as is reasonably practicable, that his design:

- ☐ eliminates foreseeable risks to the health and safety of all persons performing the construction work, cleaning the structure, or who may be affected by the work of such persons
- ☐ combats at source all unavoidable risks to the health and safety of those persons
- ☐ gives priority to measures which protect all workers over measures that protect only one person.

The Designer must ensure that these risk control measures apply to all stages in the life of the structure being designed. In this way the CDM Regulations require Designers to perform a risk assessment of their designs.

In addition, Designers must ensure, so far as is reasonably practicable, that their designs include information for contractors about the risks associated with those designs and/or the materials to be used. Having said that, the ACOP makes it clear that Designers should not dictate construction methods to those contractors.

The Designer must also co-operate with the Planning Supervisor and other Designers to enable them to comply with their health and safety obligations. This co-operation should include exchanging information about aspects of his design when necessary to avoid or reduce health and safety risks.

Finally, the Designer must inform the Planning Supervisor about aspects of his design that could present a risk of injury when interacting with the designs prepared by others.

The Planning Supervisor

The Planning Supervisor is the person responsible for co-ordinating and supervising the design phases of a construction project so as to ensure that the designs produced limit health and safety risks, so far as is reasonably practicable. To do this the Planning Supervisor must ensure that the Designers fulfil their statutory duty to conduct risk assessments of their designs.

That review by the Planning Supervisor need only be to a standard that is 'reasonable for a person in his position to take'. As such, the Planning

Supervisor is not expected to review every design or to redraw the designs, but rather conduct a review of the risk assessment procedures that the Designers have put into place. Having said that, the Planning Supervisor may wish to recommend changes to the design or the design review process—although he has no power to insist that those changes be made.

As mentioned, the Planning Supervisor has responsibility for ensuring the preparation of a pre-tender Health and Safety Plan. While he may instruct another person to prepare that Plan for him, it is the Planning Supervisor who will be held responsible for its contents.

Finally, the Planning Supervisor must ensure that the Health and Safety File is prepared for the construction project and that it is delivered to the Client at completion of the project. As there is no definition of 'completion' in the Regulations, it is advisable to define that term in the Planning Supervisor's terms of appointment.

The Principal Contractor

The Principal Contractor will be a contractor who is working on the project and who has been additionally appointed by the Client to supervise safety on the construction site.

In this way, the Principal Contractor will be expected to supervise and assess the performance of contractors, so far as is reasonably practicable, to ensure that they are complying with the health and safety rules developed for the site, and that they co-operate with one another. Similarly he must co-ordinate the interaction of contractors on site, which will include co-ordinating emergency procedures.

It is the Principal Contractor's responsibility to ensure that only 'authorized persons' are allowed on to the site. 'Authorized persons' include those with a statutory or contractual right to enter all, or part, of the site. How this is done will depend upon the type of the project; however, the Principal Contractor need only limit unauthorized access which is foreseeable. The ACOP suggests that on large remote sites, warning signs may be sufficient, while on larger sites permanent security measures may be required.

A copy of the notice sent to the HSE must be displayed on the site by the Principal Contractor in a place and condition where it can be easily seen and read by all. Generally this will be at the site entrance, at a permanent place on the perimeter or at the site office.

The Principal Contractor has the authority to *reasonably* direct the actions of contractors whenever necessary to ensure compliance with the health and safety rules developed for the site. Correspondingly, contractors must comply with those reasonable directions.

As previously mentioned, it is the Principal Contractor's responsibility to 'develop' the Plan to its final form before construction can begin and to keep it up to date once construction is underway.

The Principal Contractor must provide information on the health and safety risks on the site to contractors, who must pass it on to their employees. Finally, the Principal Contractor must ensure that all employees are consulted on the health and safety arrangements for the site. To do this, he must consult with the safety representatives and safety committees, or make other arrangements for communicating with employees in the event that safety representatives have not been appointed.

The Contractor

A Contractor is any person who carries out or manages construction work or who organizes others to carry out such work on his behalf. In this way, subcontractors, works package contractors, project managers and/or works supervisors might all be treated as Contractors for purposes of the CDM Regulations.

Specifically the Regulations oblige Contractors to:

- ☐ co-operate with the Principal Contractor
- ☐ provide the Principal Contractor with any information which might affect health and safety on the site
- ☐ inform the Principal Contractor of any injuries, accidents or dangerous occurrences on the site
- ☐ comply with the health and safety rules prepared for the site and
- ☐ provide the Principle Contractor with information he believes should be included in the Plan.

Civil and criminal liability

Breaching a health and safety law can bring with it both criminal and/or civil liability. This section will briefly consider both types of liability.

Criminal liability

Part 33 of the HSWA makes it an offence for a person to fail to discharge a statutory duty prescribed by that Act or any subsidiary legislation. The form and severity of that offence will depend on the nature of the breach.

Enforcement proceedings

The most common form of criminal enforcement action in health and safety cases is the issuance of an improvement or prohibition notice. An improvement notice can be served by an inspector if he believes there has been or may be in the future a breach of relevant legislation. The notice will require certain steps to be taken to correct that breach within a stated period, which must be at least 21 days.

A prohibition notice can be served whenever an inspector believes that a hazardous state of affairs exists which presents a real risk of serious personal injury—whether or not there has been a breach of any specific legislation. The effect of such a notice is to order the relevant activity to be stopped, either immediately or after a stated period, until certain specified steps have been taken.

For more serious breaches of health and safety laws, a criminal prosecution may be brought before a magistrates' court or a crown court. While most offences can be heard by either court, the less serious offences are dealt with in the magistrates' court, with the more serious offences dealt with by the crown court.

The maximum fine that can be imposed in the magistrates' court for breaches of the principal parts of the HSWA (Parts 2–6) is £20 000. The maximum fine that magistrates can impose for other breaches of other parts of the Act or of other statutory instruments is £5000. Magistrates can impose custodial sentences of up to six months for breaches of certain improvement notices. The crown court, on the other hand, can order an unlimited fine for any offence and prison sentences of up to two years for certain offences.

Directors' and officers' liability

Part 37 of the HSWA makes it clear that officers and directors of a company can be prosecuted for a safety offence if it can be shown that the company committed that offence 'with the consent or connivance of, or to have been attributable to any neglect on the part of any director, manager, secretary or a similar officer of the body corporate or as a person who is purporting to act in any such capacity'.

A director found guilty under Part 37 of the HSWA can be fined in the same way as the company. In addition, he can be disqualified for up to two years from serving as a company director, by virtue of the Company Directors (Disqualification) Act 1986.

Employees' duties

There are a number of duties imposed on employees under health and safety law. These include Part 7 of the HSWA which provides that every employee shall take reasonable care of his own safety and that of others. Similarly, Part 8 of the Act provides that no person shall misuse any health and safety precautions or equipment. If an employee is prosecuted, he would be subject to the same fines mentioned above.

Burden of proof

For any defendant to be convicted of a criminal offence, the prosecution must show 'beyond a reasonable doubt' that the accused committed that offence. However, the HSWA makes the prosecution's job somewhat easier in health and safety cases by transferring part of that burden of proof to the accused.

Specifically, Part 40 of the HSWA provides that the accused must show that his actions were either:

- practicable
- reasonably practicable, or
- that there was no better practical means than those used to satisfy the particular duty or requirement.

If the accused cannot satisfy this burden, the case will be considered proven against him. Obviously, this burden can make it much more difficult for defendants to defend successfully a health and safety prosecution.

Manslaughter actions

In the event that a breach of a health and safety law results in death, the responsible party could face a charge under general criminal law for manslaughter—as compared to a charge under health and safety law. 'Manslaughter' is defined as death caused by the unlawful act or gross criminal negligence of a person and is punishable by a fine.

It is possible to bring a manslaughter charge against either an individual or a company; however, prosecutions of companies for manslaughter have traditionally failed for proof problems. Having said that, in 1995 the UK saw its first successful prosecution of a company for manslaughter in *R* v. *OLL Ltd* [1995]. OLL Ltd was the company that organized the 1993 school canoeing trip which led to the death of four teenagers. Since then only one other company has been convicted of manslaughter in *R* v. *Jackson Transport (Ossett) Ltd* [1996].

It is likely that the number of corporate manslaughter convictions will rise at a much faster rate if the Law Commission's proposal for a new offence of 'corporate killing' is adopted. That offence would supplement the existing offence of corporate manslaughter, but would be much easier to prove. The proposed corporate killing offence is described in the Law Commission's paper entitled *Legislating the Criminal Code: Involuntary Manslaughter*, published in March 1996. At the time of writing it is not clear if, or when, this proposal might come into force.

Civil liability

In addition to criminal proceedings, a person who breaches health and safety laws may face a civil action for damages if that breach causes injury or damage. Typically, plaintiffs will bring actions for negligence, breach of contract and/or breach of statutory duty, which actions are summarized below.

Breach of contract

The duty to take reasonable care for the safety of an employee is one of the terms of the employment contract between an employer and an employee, whether expressly stated in the contract or implied as a matter of law.

Either way, if an employee is injured as a result of an employer's failure to take reasonable care, that employee can pursue a civil action against his employer for breach of contract. To succeed, the employee would have to show that:

- ☐ there was an employment contract with the defendant
- ☐ the defendant breached that contract by failing to take reasonable care, and
- ☐ as result of that breach, the employee was injured.

Negligence

Employers owe a common law duty of care to ensure the health and safety of their employees and others. Specifically, employers have a common law duty to ensure:

- ☐ safe and adequate plant and equipment
- ☐ safe premises and/or place of work
- ☐ competent and safe fellow workers, and
- ☐ a safe system of work.

This duty of care obliges employers to take only *reasonable* steps to prevent harm. If he fails to take reasonable care and as a result someone is injured, the injured party may be entitled to recover damages for negligence.

The courts recently extended the application of an employer's common law duty of care to his employees in *Walker* v. *Northumberland County Council* [1995]. It was held in that case that an employer owes a duty to ensure that both an employee's mental health and his physical health are not jeopardized by his employment. It was for this reason that the court in *Walker* felt fit to award damages to an employee who suffered two nervous breakdowns caused by overwork and stress.

Breach of statutory duty

The most recent weapon in plaintiffs' arsenal of civil claims is the 'breach of statutory duty' claim, which permits an injured party (typically an employee, but not necessarily) to bring an action against any person whose breach of a health and safety statute caused him injury.

To succeed in a claim for breach of statutory duty, the injured party must show that

- ☐ a duty was breached that can give rise to a civil liability
- ☐ that statutory duty is owed to the plaintiff by the defendant

- ☐ the damage suffered by the plaintiff is of a type which the statute was designed to prevent, and
- ☐ the injury was caused by the defendant's breach.

It is not generally possible to bring an action for breach of a statutory duty that relates solely to employee *welfare*; however, it is generally possible to bring such an action for breach of a duty that relates to employee safety.

For example, under the CDM Regulations a plaintiff can bring an action for breach of statutory duty only when it can be shown that the Principal Contractor breached his duty to limit access to the site or when a Client permitted construction to begin before the Plan was fully developed. A plaintiff is not entitled to bring an action for breach of statutory duty in for breaches of any other section of those Regulations.

Conclusion

Even this brief summary demonstrates the complexity of the law of health and safety. Therefore, individuals who believe that they may have health and safety responsibilities are advised to review the statutes and guidance carefully before acting.

2. Professional ethics and morals

David I. Blockley, *University of Bristol*

Introduction

Story 1

You are a civil engineer specializing in welding. You are hired to inspect some steelwork construction for a nuclear power station. You immediately find that the welding practices on the steel pipes that are to carry cooling water are substandard in a number of respects. You find that the wrong kind of electrodes are often being used. You notice that the oven drying required for some of the electrodes is not being done. You discover that many of the welders are not properly qualified and many are simply learning on the job. You raise these issues with your boss but nothing gets done and when you enquire further you are told to be quiet about it.

What would you do?

Carl Houston working in just this situation in Surrey, Virginia, USA in 1970 (Martin and Schinzinger, 1997), kept reporting to his bosses and within two months was forced to resign. Subsequently he notified every organization involved in the work with little response. Eventually he managed to get two senators from his home state of Tennessee to take up the case. An independent firm of consultants was appointed to investigate the claims and all of the deficiencies were confirmed. Houston was technically vindicated but suffered severe financial losses.

Would you have done what Carl did?

Story 2

You have just worked long hours for several days to help your firm put in a bid for a job. The work involved some fairly lengthy calculations. The bid has been signed and sealed by the boss and delivered with only minutes to spare. You go home for a well-earned rest. You wake up in a panic after only two hours of sleep—you realize you have made a mistake and there is a serious error in the calculation. You can only speculate on the consequences.

What would you do?

You could own up straightaway—and risk the sack. You could say nothing—and hope no one will notice as the calculations were rather detailed and lengthy anyway.

Mary, a new graduate engineer, was in just this dilemma and she immediately told her boss. He thanked her for telling him, looked into the situation and realized that the mistake could be easily dealt with. However, had the error gone unnoticed safety would have been threatened. Later in the year that boss came to the university, told me this story and asked if there were any more graduates like Mary who he could recruit.

Story 3

You are a chartered civil engineer and you are travelling home on the train feeling rather tired after a long day at work. You meet an old friend who is a bit of a bore. He talks on and on about this, that and the other, and you are trying to be friendly but not really listening. He begins to tell you about the home alterations he is doing—he is an amateur DIY fanatic. He is going to knock out a load-bearing wall and says he has an old steel UB he found somewhere. He asks you whether you think it will be OK. Without really realizing it you say yes—just to shut him up.

Some time after that casual conversation you get a letter from a solicitor informing you that you are to be sued for giving negligent advice to your old friend because the beam he used was not strong enough and the wall collapsed.

How would you react?

If you were indignant and denied it was anything to do with you then that is understandable but unfortunately wrong. You should know that you have a duty of care as a chartered engineer not to give negligent advice. Any of your peers, with a simple calculation, would be able to show that the UB suggested by your friend was too small.

Is safety an ethical or moral issue?

Science and engineering are typically portrayed to the general public as 'objective' and value-free. The image that this creates in many people's minds gives the impression that scientists and engineers only deal with matters of fact so that 'subjective' opinion and judgement have no place. Engineering theory is true or false they believe, so the feelings and emotions of individuals have nothing to do with the job that they do—if anything goes wrong then someone has made a mistake and should be blamed.

Of course the reality is different, as the simple stories above make clear. Engineering is about making decisions under conditions of extreme uncertainty. Not only do engineers have to make very difficult decisions, those decisions may involve ethical dilemmas where the values of the individuals, or the society of which they are a part, have to be considered.

So let us examine first some of the central ideas— the values, objective and subjective knowledge, morals and ethics.

Values

We all make decisions based on our *preferences*. Preferences are based on *values* which are the *worth* we ascribe to something.

The goal of science is to produce *true and precise* knowledge for explanation and prediction. The goal of engineering practice is to produce systems of artefacts and people that are *fit for purpose*, in other words, they perform in the manner required. Thus the value for science is truth and precision but the value for engineering practice is fitness for purpose.

Quality is degree of excellence, which is the state of having pre-eminence or having the highest value. Therefore quality for science is truth and precision. Quality for engineering practice is fitness for purpose.

Rigour in science is the way in which truth and precision are achieved. Rigour in engineering practice is the way in which fitness for purpose is achieved.

Objective and subjective knowledge

Clearly quality in engineering is multifaceted and must be defined in a manner which is appropriate for the problem to be solved. Some values are *subjective* (i.e. entirely personal), for example, I prefer not to have a pain in my stomach, but is my pain worse than your pain?

Some values are *intersubjective* (i.e. they are shared experiences; for example, we both see the same building). These may be values that we cannot measure easily—but about which we can form judgements (e.g. environmental impacts such as beauty).

There are facets of quality which we can measure or establish with some degree of dependability (e.g. if we value tall buildings then we can specify and measure their height). Engineers tend to call these measures *objective*.

Karl Popper (1976) described three worlds of knowledge. World 1 is, according to him, the world 'out there' and can only be reached via our subjective minds, which he named world 2. However, it is Popper's idea of world 3 that is interesting. This is the world of our collective knowledge—it may be helpful to think of it as everything contained in all libraries. This knowledge is greater than any single mind can know and so has an objective existence outside each of us. It is contained in the form of world 1 matter (i.e. in books and papers and on the Internet) but it is not necessarily true; indeed, much of it is clearly false (e.g. works of fiction). Almost every great idea of human existence has eventually been shown to be false. This is discussed further by Blockley (1980) and Rorty (1982).

Engineers and all practical decision makers are not interested primarily in *true* knowledge; rather, they are interested in *dependable* knowledge. When an engineer makes a high level statement such as 'this structure is safe', or more detailed statements such as 'the yield strength of this steel is 200 kN/mm^2' or 'the safety factor on this structure is 1·53', then he or she does not really believe the statement to be absolutely true because he or she is aware of the inherent uncertainty associated with them. Unfortunately other people, particularly lay people, often take these statements at face value. If decisions are taken on the basis of that accuracy which show them to be mistaken, then friction and further misunderstandings can occur. It is important that the uncertainty associated with any piece of information is appreciated and if possible estimated in some way. For a discussion of the idea of dependable information see Blockley (1980).

Engineers have a legal and moral *duty of care* to deliver quality. The duty of care is a *responsibility* to act *reasonably* with respect to a number of constituencies—to oneself, family, client, society at large and the physical environment. The taking of responsibility implies not that one has earned the right to be right or even nearly right, but that one has taken what precautions one can reasonably be expected to take against being wrong. Thus engineering is a responsible decision-making activity based on specific values—one of which is to act responsibly by obtaining dependable information.

Morals and ethics

Morals are concerned with right and wrong—with what *ought to be* rather than what is. They are about defining what is *good*. Ethics is the discipline of moral conduct—the word suggests a set of standards by which a particular group or community decides to regulate its behaviour—to distinguish what is acceptable in pursuit of their aims and what is not. In ordinary discourse the words may be used interchangeably; for example, we may speak of moral conduct or ethical conduct, and we may speak of moral philosophy as ethics. We may speak of medical ethics or business ethics. The subject of engineering ethics is beginning to be more widely discussed. Of course there are standards of morality that are related not simply to particular roles but to all—for example, Christian moral teaching.

Engineering ethics involves normative enquiries aimed at identifying and justifying the morally desirable norms and standards that ought to guide us. Normative questions ask 'What ought to be?' and 'What is good?'. Questions are those such as 'When should an engineer blow the whistle on dubious practice? or 'Whose values should be primary on a large contract in a developing country?'.

So what are the moral or ethical principles that lie at the root of engineering safety?

Ethical principles

The 'golden rule' is the cornerstone of religious understanding. Here are some variations on it (Moses, 1989):

- ☐ Do unto others as you would have them do unto you—*Christianity*
- ☐ Do unto all men as you would wish to have done unto you, and reject for others what you would reject for yourselves—*Islam*
- ☐ Hurt not others with that which pains yourself—*Buddhism*
- ☐ This is the sum of all true righteousness—
  Treat others, as thou wouldst thyself be treated.
  Do nothing to thy neighbour, which hereafter
  Thou wouldst not have thy neighbour do to thee.—*Hinduism*

In terms of engineering practice a basic rule might be (variations of which appear in many institutional codes of conduct): 'the engineer shall apply his specialized knowledge and skill at all times in the public interest, with honesty, integrity and honour.'

Ethics is about managing differences and making choices about individual, family, work and societal issues. It is about deciding what are criteria for deciding what is best in general and in particular what is best for one person/group over another person/group.

So are ethics relative or absolute? Philosophers have tried to elucidate general principles that seem absolute but without success. It should not be inferred from this, however, that ethics are relative and hence 'anything goes' since there is a hard core of agreed principles that are defined as 'good' in all societies. The problem is that it always seems possible to find counter-arguments for any general rule which places one value over another. For example, while being honest is a very important principle, there are circumstances where it can be argued that it is best to 'tell a white lie'. Thus it is very difficult to give absolute guidance on the values to be used in a particular situation. In the main the issues are unproblematic, but at the extreme there are very difficult choices to be made, as illustrated by the stories at the start of this chapter and by the moral dilemmas described at the end of it.

Some of the important values for engineers are:

- ☐ truth, honesty, trust
- ☐ respect for others and the environment
- ☐ fairness
- ☐ openness
- ☐ competence

- ☐ sustainability, balance and harmony.

You may be able to think of others that are particularly important to you.

It is important that engineers, in exercising their responsibilities, should:

- ☐ be clear about what their own values are
- ☐ be informed about the projects in which they are involved
- ☐ think about the consequences of what they do and in particular try to anticipate unintended and unwanted consequences
- ☐ be up to date in professional skills
- ☐ act professionally only within their area of competence
- ☐ keep health, safety and public welfare as paramount
- ☐ communicate openly with the public about technological developments
- ☐ be honest
- ☐ disclose circumstances where there may be a conflict of interest
- ☐ neither offer nor accept bribes
- ☐ treat all others fairly in respect of race, religion, sex, age, ethnic background or disability
- ☐ help colleagues promote professional growth.

Institutional codes of ethics and culture

Engineering ethics has perhaps been rather too narrowly conceived as consisting only of codes and guidelines from professional societies and institutions. All of the major professional engineering institutions have a 'code of conduct' which they expect members to follow. These vary in detail but most of them address the values set out previously. The advice with respect to safety is sound and works well for normal situations but is not really helpful for sorting out difficult dilemmas when there is a clash between basic values.

Of course, the context of our time tends to shape ethical behaviour. There has been in the recent past perhaps a concentration on short-term financial gain—courage is needed to speak out against it. There has been perhaps, for a few, some confusion about right and wrong—some people have turned a blind eye to 'goods off the back of a lorry' or to the black economy. Pressures of consumption have, for many, overridden worries about pollution.

It seems that regulatory frameworks and codes of practice are not sufficient to force the integration of ethics into professional practice. There is an urgent need to educate staff and students. All students should be aware of the requirements of good citizenship. Specific details of exercising proper conduct with respect to safety will require continuing education (e.g. in new regulations such as the recent introduction of the Construction (Design and Management) (CDM) Regulations 1994) and in updating ideas with respect to the latest thinking in avoiding unsafe practice. We really do not understand the nature of an ethical culture.

Safety culture

There has been much debate in recent years about the nature of a safety culture (Blockley, 1996). The common-sense view of a 'culture' could be summed up in the phrase 'the way we do things around here'. There may be an agreed part (that which we take for granted) and a questioned part (which is problematic). Too often compliance cultures and efficacy rule the day. We need to be aware that the changing context and concerns over, for example, environmental issues shape the perceptions of young people.

As noted earlier, engineering is about creating systems of artefacts to satisfy some function or purpose. Quality is considered by some engineers as relating only to function and therefore reliability. However, if quality is defined in general terms as 'fitness for purpose', then clearly the functional purpose cannot be served if the artefact is unsafe. Safety is therefore a necessary condition for quality. The provision of both safety and reliability is therefore one part of the management of quality. The construction industry has a very poor safety record. The connection between lack of safety and real cost has now been directly demonstrated (Health and Safety Executive, 1993).

In 1978 Turner (in the first edition of the book recently republished as Turner and Pidgeon (1996)) argued that the potential for unintended and unwanted consequences of human actions can develop over time. He showed that large scale accidents usually have multiple preconditions with factors which accumulate over a considerable period of time (called an *incubation period*). For example, events may be unnoticed or misunderstood because of the wrong assumptions about their significance. Dangerous preconditions may be unnoticed because of the difficulty of handling information in complex situations. There may be uncertainty about how to deal with formal violations of safety regulations. When things do start to go wrong the outcomes are typically worse because people tend to minimize the danger as it emerges, or to believe that the failure will not happen. As a result, events accumulate to increase the predisposition to failure. The 'size' of the trigger event (e.g. a high wind, an earthquake or a simple human error) which releases the energy pent-up in the system is not the only important cause of the accident; rather, one of the main tasks following an accident is to identify the preconditions. These preconditions are the hazard, and they represent the developing 'potential' for failures and accidents.

Blockley (1996) has argued that if hazard incubation is a process then the current interest in business process re-engineering is helpful to the business of hazard management. However, the most effective way to improve the safety performance of a construction company is to change its safety culture. The basic problem is the traditional view that safety and profitability are in direct conflict.

The ingredients of a good safety culture are similar to those of a quality culture. The following acronym 'practical learning' illustrates some of the important factors that are suggested as being present in a company with a good safety culture (Blockley, 1996).

- P *point of view*—remember we all have one and should respect those of others
- R ensure *responsibilities* and accountabilities are clearly defined
- A encourage positive *attitudes*
- C have good *communications*—both formal and informal
- T ensure the *technology* is appropriate
- I have clear *intentions or objectives*
- C show a *caring* approach to safety (especially needed in senior management)
- A keep all workers *aware* of quality and safety issues
- L encourage a *learning* approach with an open mind—see failure not as a cause for blame but as an opportunity to learn

- L encourage active *listening*—talk to people not at them
- E give *encouragement*—give praise at every genuine opportunity
- A keep *auditing* by testing for hazard—measure and check through performance measures
- R keep a firm grip on *reality*—do not allow yourself to be deluded, keep your feet on the ground and face facts rather than shy away from them
- N encourage *new ideas*—help people to be creative in 'owning' and hence changing processes to clear objectives
- I make *information* as clear and unambiguous as possible using IT wherever appropriate
- N have *no 'yes men'*—truth is found from honest disagreement among friends
- G be *genuine*, honest and open.

These characteristics of a good safety culture depend clearly upon the ethical stance of the players taking part in the construction process.

Classes of dilemmas

So we conclude that there seem to be basic ethical principles that we should adopt, but it is difficult to decide how to prioritize them. We may gain insight into how this might be done by examining classes of ethical dilemmas.

Of the three stories at the start of this chapter the first two are commonly seen as examples of 'blowing the whistle' and the third is a straightforward example of exercising a responsible 'duty of care' at all times—even on a train! In the first story the central difficulty is in balancing a responsibility to society with a responsibility to yourself and your family. In the second story the essential issue is 'Do you own up when you have made a mistake?'. Do you risk not being found out when something awful happens or alternatively could you live with yourself if someone was killed because you did not point out your error in time?

There are clearly many ways in which one could classify such ethical dilemmas—what follows is one suggestion. The five classes are 'blowing the whistle', 'producing unethical artefacts or systems', 'not exercising proper responsibility to future generations', 'succumbing to a dictatorship of the uninformed' and 'losing the diversity of systems?'. I will describe each one in turn briefly.

Blowing the whistle

The dilemma here is whether to report to some relevant person some moral corruption where that relevant person is able to register a response—if only in protest. A central difficulty is exemplified in story 1, where the interests of oneself and one's family have to be offset against the interests of the wider community. Clearly the nature of the moral corruption is important—there may be a clear distinction in many people's minds between criminal behaviour such as theft and some slight bending of the rules to 'fiddle a tax return'. So when is whistle blowing morally permissible? Is it ever obligatory? When is it disloyal?

There are several factors with respect to safety that should be considered, some of which are as follows:

(a) Is there a serious threat to the safety of the public?
(b) Has no satisfaction been obtained from one's immediate bosses or, for whistle blowing to people outside of the organization, from higher level bosses including the board of directors?
(c) Is convincing documented evidence available?
(d) Is there evidence that by blowing the whistle the safety of the public will be maintained?
(e) Will responsibilities to one's family not be unacceptably harmed?

You may well be able to think of other factors.

In the final analysis only the individual engineer (perhaps after discussion with others) can make a decision about whether or not to 'blow the whistle'. However, each one of us has a responsibility to consider our values well and to balance them against each other in a particular situation; it is a judgement, but it must be a responsible one, taking a proper duty of care. The expression 'Don't blame me I was just doing my job' has no place in the life of a responsible engineer.

Producing unethical artefacts or systems

The dilemma here is represented by the question 'If you had been asked to design or build the Nazi gas chambers would you have done so?' Doubtless all of us would say that we would not. But what is an 'unethical artefact or system'? Some would argue that the new road bypassing the town is unethical because it requires some trees to be cut down—the 'protesters' then try to stop its construction by chaining themselves to the trees or by digging dangerous tunnels.

The problem is that our view of what is unethical about any particular artefact or system depends on our point of view. In a democracy, points of view can vary enormously. The legal trade in arms is considered by many to be unethical and by others to be the way to prevent war between nations (e.g. the argument to 'balance' nuclear arms between East and West—which has more or less gone since the fall of the Soviet Bloc). The way to solve such dilemmas derives from both personal and political values, but most people are content to allow the democratic procedures of our society to resolve them. However, those procedures can often be criticized and then it is a matter of personal choice. Nevertheless, the engineer has a duty of care to consider these dilemmas responsibly. The expression 'If I don't do it someone else will' has no place in the life of a responsible engineer.

Responsibility to future generations

This is a specific set of dilemmas that could be considered as a subset of those concerning unethical products or systems. We have a particular responsibility to our children and their children that must affect our decision making. Issues such as global warming, pollution and long term energy sources are enormously difficult in this respect but represent long-term threats to the safety of people. The responsible engineer will always consider the effects of what he or she is doing on the unborn.

The dictatorship of the uninformed

There is also a general problem which could also be considered as a subset of the unethical artefact or systems but which should be singled out for special attention. There has been, in recent years, a loss of confidence in the idea of an 'expert'. A general questioning of authority, scandals in the City, corrupt lawyers who end up in jail and large-scale engineering failures have all contributed to the loss of faith in the 'expert'. Professional people have to grapple with complex problems within complex systems and have to make judgements which lay people have to take on trust. Trust is a very fragile commodity which once broken is very difficult to rebuild. Once a few professionals are seen to be untrustworthy then unfortunately others are dragged in, with a resultant reduction in the confidence in the profession as a whole. This loss of confidence leads to a dangerous 'relativism' where everyone's ideas are seen to be equal, no matter what the skills or expertise that person possesses. If major decisions are made this way because they are seen to be democratic, it is likely that the democratic society itself will be damaged. A healthy society has to be built on trust so that decisions are made by those best informed to do so and the rest of us feel able to trust them.

Diversity of systems

Walk into some of the shopping malls in Kuala Lumpur and you could be in the USA or, indeed, the UK or anywhere in the world. The globalization of our economies means that there is a loss of diversity in the technologies that we bring to different parts of the world. If one takes a perspective that engineering develops in an evolutionary way, similar to natural systems, then this loss of diversity is a serious threat to our future ability to find novel solutions to our problems. It is ethical to encourage diverse solutions to technical problems which take into account much more explicitly the local cultures of the countries concerned.

Conclusions

(a) Science and engineering are typically portrayed to the general public as 'objective' and value-free. However, engineering decisions may involve ethical dilemmas where the values of the individuals, or the society of which they are a part, have to be considered.

(b) *Values* are the *worth* we ascribe to something. Quality for science is truth and precision. Quality for engineering practice is fitness for purpose.

(c) Engineers have a legal and moral *duty of care* to deliver quality. The duty of care is a *responsibility* to act *reasonably*.

(d) The 'golden rule' is 'Do unto others as you would have them do unto you'. Although there are basic ethical principles, there seems to be no universally acceptable way of prioritizing them in all circumstances. Classes of ethical dilemmas may be a way of making clearer some of the choices.

(e) Ethics is about managing differences and making choices. Engineering ethics should be about helping engineers to be clear about their own value systems.

(f) Regulatory frameworks and codes of practice are not sufficient to force the integration of ethics into professional practice. There is an urgent need to educate staff and students. Specific details of exercising proper conduct with respect to safety will require continuing education (e.g. in new regulations such as the recent introduction of the CDM Regulations) and in updating ideas with respect to the latest thinking in avoiding unsafe practice.
(g) The attributes of a 'good' safety culture have been suggested through the acronym 'practical learning'.
(h) Five classes of ethical dilemma have been suggested: 'blowing the whistle', 'producing unethical artefacts or systems', 'not exercising proper responsibility to future generations', 'succumbing to a dictatorship of the uninformed' and 'losing the diversity of systems'.

Bibliography

Blockley, D.I. (1980). *The Nature of Structural Design and Safety.* Ellis Horwood, Chichester.
Blockley, D.I. (1996). Business process engineering for safety. *15th Congress IABSE, Copenhagen*, 665–676.
Health and Safety Executive (1993). *The Costs of Accidents at Work.* HMSO, London.
Martin, M.W. and Schinzinger, R. (1997). *Ethics in Engineering.* McGraw-Hill, New York.
Moses, J. (1989). *Oneness.* Fawcett Columbine, New York.
Popper, K. (1976). *Conjectures and Refutations.* Routledge and Kegan Paul, London.
Rorty, R. (1982). *Consequences of Pragmatism.* Harvester-Wheatsheaf, Hemel Hempstead.
Turner, B.A. and Pidgeon, N.F. (1996). *Manmade Disasters*, 2nd edn. Butterworth-Heinemann, Oxford.

3. The CDM Regulations: keeping an air of realism

Martin J. Barnard, *Symonds Group, East Grinstead*

Introduction

Much has been debated about the way the Construction (Design and Management) (CDM) Regulations 1994 have been implemented since they came into force in March 1995. We have progressed a long way from the gloomy pictures painted of designers without professional indemnity and prosecution *en masse* by the Health and Safety Executive (HSE). In fact we have seen a pragmatic approach by the latter entirely in keeping with the 'sense of proportion' they have avidly promoted for the industry. We have found that innovative design is not stifled; indeed, in many ways it has been enhanced. A full description of the duties under the CDM Regulations is given elsewhere in this book. This chapter attempts to take a step back from the intensity of the theoretical debate and view the CDM Regulations from the author's practical involvement with numerous high profile clients involving projects whose total construction value exceeds £1000 million.

The good bits

Integration of health and safety

It is in this area that the greatest strength of the CDM Regulations lies. They provide the means to integrate health and safety into projects. They force organizations away from traditional 'bolt-on' safety, seen as being someone-else's problem. Many had in fact been dealing with safety reasonably well before the CDM Regulations; they simply did not recognize it as such. Increasing familiarity has encouraged the key players to deal with health and safety in a proactive manner as a significant and important part of the project (Fig. 1). Indeed, I have seen occasions where some have gone 'over the top' in their effort to do the right thing. That is part of the learning process which encourages me to say that the CDM Regulations are now starting to work as intended—as an integral part of a construction project.

Clarity of duties

Whatever views are held about the CDM Regulations, there can be no doubt that they have substantially cleared the muddied waters of 'who is

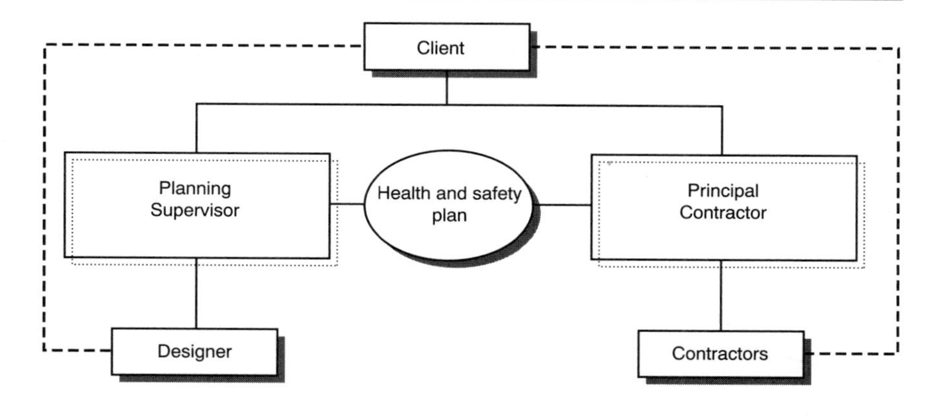

Fig. 1. CDM Regulations duty holders

responsible for what?' which existed previously. The health and safety 'cake' is cut up in a very clear way. While duty holders may not like the slice they have, they can be in no doubt as to the extent of their duties and what others in the project team expect of them.

Emphasis on pre-planning

The ability to stand back occasionally and consider the aim of the CDM Regulations is of particular value. At the sharp end, in other words during the construction/maintenance work, the aim should be to get *the right people in the right place at the right time doing the right job.* If that can be achieved then the CDM Regulations are satisfied. For that to happen, emphasis must be placed on the effective planning for health and safety, often at the design stage. The principles that the CDM Regulations promote are the same as those recognized as fundamental to good project management; in other words, good planning and preparation leads to effective delivery. I see clear signs that pre-planning for health and safety is being adopted to increasingly good effect.

Weeding out the cowboys

Quietly but surely this process is happening. It is being led by clients who increasingly recognize the influence they can exert in such areas. Most clients look astounded when I ask them if a contractor has ever refused to work for them. The reality is that if a client shows that he clearly intends to enforce high standards of health and safety, those who serve him will comply. Half-hearted attempts by clients in the past have exacerbated this problem and encouraged the cowboys. Now they either comply or go out of business. Indeed, some clients are thriving on the acquisition of a legal lever to control their contractors more effectively.

Health and Safety Plan

The Health and Safety Plan has often been misunderstood and occasionally used in totally the wrong way. If used correctly it becomes the core element for effective communication and control of health and safety both before and during construction work. For larger organizations the use of a 'generic plan framework' becomes the means by which they can gain consistency of approach and the commercial advantage which comes from their numerous project teams being familiar with the purpose and use of the Plan. This obviously needs to be done in a way which does not tell others how to do their work while still gaining from the avoidance of the continual 'reinvention of the wheel' by separate project teams. I believe the content of a health and safety plan should be simple and straightforward. It is for that reason that I have developed a rule that *the usefulness of the plan is inversely proportional to its size.*

It is in this area perhaps more than most that the Planning Supervisor earns his credibility. It should be recognized that much of the Plan is information giving. I believe the Plan should also set 'performance standards' for health and safety which can be treated in the same way as a technical specification. Indeed, a well-written Plan forms the baseline for judgements to be made in relation to resource, development of the Plan, Regulation 10 approval, etc. Many believe that only 'unusual' hazards should be indicated in the Plan and routine, obvious ones excluded. I see it the other way round. The Plan should address the principal hazards, which I define as those which cause most harm on a most frequent basis. In construction this is often falls from height or being run over. If we limit ourselves to the 'unusual', many projects will have no substance to the Plan.

Health and Safety File

Along with many others, I have been pondering on the most effective way in which the Health and Safety File can be used. The concept of providing health and safety information to those who follow is commendable. How you achieve that in practice needs more consideration. My current thoughts lead me to highlight the practical problems in accessing particular elements of the information contained in what I term the 'minimalist' approach to the File containing the information advocated in Annex 5 of the CDM Approved Code of Practice (ACOP). While satisfactory for large future projects which will utilize much of the information in the File, the minimalist File is likely to be more frustrating in the prompt provision of selective information for a smaller maintenance task. Of course a small CDM Regulations project will generate a small File, where access to individual elements is immediate and obvious. However, larger projects need to go beyond the minimalist approach if they are to be effective in all areas. I increasingly advocate using a comprehensive file-indexing system which can readily identify the information available for selected maintenance activities. While costing more to set up, it will repay itself several times over given the frequency of maintenance activities. For it to be effective we need to go back to basics and question what we are trying to do. With that in mind, I advocate close liaison with the end-user. Some think that the view that the end-user should be consulted is revolutionary; I see it as fundamental or why bother doing it? What is important is that we can distinguish between the end-user's 'wants' and actual 'needs'. In that way, an efficient and effective file can be produced.

Planning Supervisor

Before we progress to consider the Planning Supervisor role in practice, it is necessary to address the formality of the role. I have always felt very strongly that it should be a corporate appointment and not an individual. It is pleasing to note that the majority in the industry now follow that line.

The next stage is to consider who is to play the role. Virtually all clients see themselves as dependent and reliant on the Planning Supervisor. They seek reassurance that the Planning Supervisor will act in an independent and impartial way. For those reasons I offer up the following as a client's order of preference for Planning Supervisor:

(a) independent of project team
(b) part of the client's organization
(c) part of the designer's team
(d) part of the contractor's team.

Having covered the 'corporate' concept above, it is true to say that the day-to-day function of the Planning Supervisor will fall to an individual. It may be that he/she in turn calls on other expertise as part of the 'team'. What level of expertise should that individual demonstrate? Based on my recent experiences, I regard the following as an order of preference.

(a) Equal experience in construction and safety

This to my mind is the ideal. At present they are somewhat of a rarity and will remain so for some time to come. They will be chartered in engineering or similar, and be a Member of the Institution of Occupational Safety & Health

(MIOSH). As a consequence they will be restricted to high-profile work or in a supervisory/overseeing role with others working to their instructions.

(b) Main experience in safety, secondary in construction
Much of the role of the Planning Supervisor will be that of 'devil's advocate', for example asking critical questions of the design team. To do so effectively, he/she will need credibility with the construction professionals. That will come from 'speaking the language' and not being 'baffled by science'. It is not necessary for them to know how to design; their impact will come in their insight into the consequence. They will have MIOSH status. They will be able to operate as the Planning Supervisor on most projects and will compliment those in (a) above on the major ones.

(c) Main experience in construction; secondary in safety
For persons to be competent in safety issues they need regular experience of the subject. It is interesting to note that persons in this category sometimes fail to recognize the gap between what they think they know and that which they actually do know. They invariably need to work from the first principles of safety and must avoid too readily accepting a design as correct where, as Planning Supervisor, they should be testing the conception behind the design. They will have chartered status in engineering or similar, but no formal safety qualifications. They will have limited ability to be Planning Supervisor, but will be an effective supplement to those in (a) and (b) above.

(d) Main experience in construction; crash course in safety
Short-duration training courses have been identified as a means of converting construction professionals into instant Planning Supervisors. I believe that such an approach is flawed and dangerous. Such individuals need time, at least a year, of constant involvement in safety to become proficient. Until that level is reached, close supervision is needed by category (b) at least and preferably category (a). While they may have chartered status in a construction discipline, it will take a number of years before they become remotely near to MIOSH status.

(e) Experience in either construction or safety but not both
I believe that a person in this category is unsuitable to be Planning Supervisor on anything other than the most straightforward of projects, and only then after receiving relevant health and safety training and having ready access to a health and safety specialist for guidance.

It is essential that there is a knowledge of design and construction as well as of health and safety. It follows that in practice a combined effort is needed; in other words, and individual in group (b) above combining with one in group (c) to produce the 'planning supervisor' output.

Who is competent to do design work?

The detailed tests of competence are given in paragraphs 39 and 40 of the CDM ACOP. Design under the CDM Regulations is a wide concept, embracing as it does drawings, design details, specifications and bills of quantities, including specification of articles or substances. So not only architects and engineers are designers; quantity surveyors and anyone who specifies a material will also fall into the definition. However, for most of the 'design' of a building, the architect together with the structural and services engineers are the designers.

There is no population of designers trained in health and safety at work. While very knowledgeable about building services, some may have little or no knowledge of health and safety problems faced by those installing and maintaining such services. To be judged competent, designers should be able to produce designs that are 'buildable' and safe to maintain. This is not to say that designs that are easy to build and maintain need be pedestrian and unimaginative. Exciting innovative design is still just as possible under the CDM Regulations but must encompass more than appearance and deal positively with health and safety. In my view the CDM Regulations will not stifle innovative design but will encourage it.

We need to see 'competence' as a backward-looking exercise asking 'do they have the expertise to design for safety?' On the other hand, 'resource' is forward-looking and asks 'how will they put that expertise into practice'. Those organizations which can effectively separate the two will, in my view, be best placed to deal with safe design in an efficient way.

How far do designers need to go?

All the requirements on designers in the CDM Regulations are in Regulation 13. For completeness, the key design considerations are (subject to the test of being reasonably practicable) the need to:

- avoid foreseeable risks to the health and safety of those carrying out or affected by construction or cleaning work
- combat at source risks to health and safety for the above
- give priority to 'collective' health and safety measures
- include adequate information on aspects of the structure or materials used which might affect health and safety.

The above leads to risk assessment, which is often misunderstood and inefficiently applied. For us to be successful in the future, we need to be clearer in our thinking on the subject. The purpose of risk assessment is to indicate to designers the potential effect of their design on the health and safety of workers and others. As a consequence, designers will be able to judge the weight they should give to health and safety and whether, on balance, the design can be left unchanged or should be altered to reduce the health and safety risk. What is 'reasonably practicable' (which also involves cost and other design goals) becomes part of the designer's judgement.

A precise estimate of risk is not required; it would be too time consuming in practice and, in any case, lack of data often makes it impossible. There is a variety of risk assessment methods, ranging from the crudely 'qualitative' to the sophisticated 'quantitative'. Any method chosen will, to some degree, be subjective and arbitrary but, nevertheless, can prove useful provided it is appropriate for its purpose and its limitations are recognized. I advocate a simple but effective approach. For example, three categories of severity can be assumed, as follows:

High fatality, major injuries or illness causing long-term disability
Medium injury or illness causing short-term disability
Low other injury or illness.

Since the primary concern for designers is what they can do to eliminate hazards to reduce risk, the measures which contractors can taken on their own behalf to protect their workers (e.g. temporary edge protection, personal protective equipment) should be discounted by designers at this stage. Even though contractors could control risks which particular hazards give rise to by the application of well known precautions, the designers must still give consideration to how hazards can be eliminated and risks reduced. Designers will have to analyse the likely method(s) of construction and/or maintenance etc. to be able to make a judgement as to the likelihood that harm will occur. Designers need to consider whether hazard and worker will coincide; how many workers, how often, for how long. Again only a crude, qualitative judgement can, and need, be made:

High certain or near certain to occur
Medium reasonably likely to occur
Low very seldom or never occurs.

The product of the elements will give some measure of the assessed risk which, in turn, can be seen as exerting a pressure on designers to take action. Clearly, a 'high' × 'high' risk exerts a very high degree of pressure, a 'low' × 'low' virtually none. Designers may conclude that design alteration is not practicable, but they should be prepared to justify their choice in the light of the particular risk assessment. An example is given at Appendix 1.

Regulation 13 does not require designers to produce designs that will be totally and absolutely safe to build nor does it require a designer to tell a contractor how to do his work. A staged approach is required in applying the principles of the Regulation. Some argue that only 'unusual' hazards should be considered by designers and routine, with the obvious ones excluded. I see it the other way round. Designers should address the principal hazards, which I define as those which cause most harm on a most frequent basis. This is often falls from height or being struck by plant or materials. If we limit ourselves to the 'unusual', many projects will have no substance in relation to health and safety.

The bad bits

Bad Planning Supervisors

At various times I have been unkind to those Planning Supervisors who are trying to do the job well, by raising my concerns about the large number who claim the title but fall pitifully short of the standards. While acknowledging the growing ability among Planning Supervisors, I see increasing frustration among clients who see themselves paying large fees for little obvious return. That cannot continue without the whole concept of the CDM Regulations falling into disrepute. It is in this area that the HSE needs to get tough. Once it starts to sort out the inadequate and incompetent organizations, then the true benefits of the Planning Supervisor will be clear even to the doubters.

Testing of competence and resource

This is one of the areas which causes large amounts of paperwork and frustration among the CDM Regulations players. Standard competency questionnaires are sent out and standard replies are received back. Invariably they do not match, which causes the enquirer to view the reply as evasive and unhelpful. In turn, the busy contractor bidding for numerous projects with often a low success rate simply does not have the time to tailor the reply. If an organization was competent last month then surely it will be competent next month. Why go around the competence circuit over and over again? I defy anyone to establish that a large, well-established organization is incompetent using a questionnaire format. By definition it is a paperwork exercise which elicits paper answers in the form of theoretical manuals which are to be found in abundance. I passionately believe that we should move away from testing of competency by paperwork and allow performance on projects to be the true test. In that way, practical ability is the test of competence rather than paperwork. In my view testing of competence by questionnaire should be used less and less in favour of 'end-of-term' reporting.

I believe there is a gap between what construction professionals say they do in relation to safety issues and that which is actually done. While this has an element of 'competency' in it, we also need to consider such a shortfall in the context of 'adequate resource'. The continued credibility of the CDM Regulations will be significantly enhanced if we can have a clearer understanding of competence and resource. Currently they are confused with each other. We need to see 'competence' as a backward-looking exercise asking 'do they have the expertise to design for safety?'. On the other hand, 'resource' is forward-looking and asks 'how will they put that expertise into practice?'. Those organizations which can effectively separate the two will, in my view, be best placed to deal with the CDM Regulations in an efficient way.

Adequate resource

I have indicated above that we see the test of adequate resource as being particularly important in the CDM Regulations. While not part of this presentation, I see this aspect as being the most difficult but most relevant exercise to be undertaken. In short, I see it as the second stage of a two-stage exercise (competence being the first) which focuses on the most likely successful organization rather than those who will not be charged with the task.

Judgement of resource can be as simple or as complicated as you wish to make it. Invariably someone has to make a judgement as to what is needed and how it will be achieved. It is at this fundamental stage that errors can be made. If those carrying out the initial review of resource are unsuited to the task then the whole exercise may be doomed to failure. Assessment of resource is critical to the Regulations and should only be done by those with the appropriate knowledge.

It has been my experience that the 'willing amateur' often lacks the ability to stand back and look with fresh eyes at what is required. He is often intimately involved in the work and lacks the ability to go back to stage 1 in the review; in other words, he takes certain things for granted. It is often in that first stage that the error exists. Another problem for the 'willing amateur' is the narrowness of his experience. He simply cannot recognize potential hazards and assimilate the significance to his activities. Even where the assessments are fulfilled, we move into what is probably the greatest 'danger': overstating that which can be achieved.

It is human nature to write what we think. In wanting to do the right thing in safety, we tend to think in superlatives—'a particular feature must be in place on each occasion'. It is sometimes impossible to achieve such a high level and unlikely to be 'reasonably practicable' to do so. I have seen several instances in the past where organizations have fallen foul of impossible standards and paid a high price as a consequence.

Confusing definitions

Those who talk about the CDM Regulations rather than complying with them have raised theoretical problems in all sorts of areas. We have noticeably moved a considerable distance from the early scare stores of the CDM Regulations being unworkable, designers not getting professional indemnity cover, etc. There are, however, a number of issues which remain confusing to practitioners.

Demolition

Much debate surrounds this definition. Often it degenerates into the puerile world of 'is drilling a hole in a wall defined as demolition?'. Even the greatest cynic of the CDM Regulations would recognize that this is not what the CDM Regulations are about. As time has gone by, I have become increasingly convinced that the demolition work which the CDM Regulations wish to address will invariably be captured by the five-person rule. While not yet able to research the point, I wonder how many have been injured in demolition jobs which had less than five persons involved.

Roofwork

While not a point of definition, the difficulties which surround 'demolition' cause me to wonder if the CDM Regulations have addressed the right issue in respect of short-duration, high-risk work. The workforce at risk in short-term roofwork appears to me to be significantly larger than in demolition. Many roofing jobs will have less than five people involved and will therefore escape the CDM Regulations. The definition of roofwork is far easier to determine than demolition appears to be. Perhaps there is an argument to add roofwork to demolition as being 'CDM-able'—pronounced 'condamable'—irrespective of size of job.

Keeping the Health and Safety File up to date

The use of the Health and Safety File as a source of information for future work is widely acclaimed as a positive feature of the CDM Regulations. The absence of any legal duty to keep a File up to date causes many to wonder if the full potential of the File is being exploited. While not underestimating the task of incorporating non-condamable 'tinkering' work into the File, many wish to see efforts made to try and do so. At the moment the pressure for up-to-date Files is coming from the commercial quarter in that the value of a property will be adversely affected by the absence of an up-to-date File. The development of this issue will remain one of the most interesting in the coming year.

Overseeing the construction phase

The CDM ACOP is clear in saying that neither client nor Planning Supervisor need get involved in overseeing how the construction phase plan is used. I continually find it hard to explain to a lay person why this is so during what is the most important time for the CDM Regulations—during the construction work, when most people are at risk. Perhaps the answer lies in the yet to be fully explored concepts of the new Construction (Health, Safety and Welfare) Regulations in respect of the duties of those involved in the project.

Local authority exemption clause

This must be the biggest 'red herring' in the CDM Regulations which has caused time and effort to be wasted in vain attempts to avoid having to comply with the Regulations. The presence of this exemption is based on history and, no doubt, political expediency. The industry's attempts to interpret the clause in a meaningful way have moved from initial purist intrigue to frustration at its continued mis-application. I hope in time it will revert to its true status as a minor issue under the CDM Regulations.

Definition of 'structure'—is it really a can of worms?

The definition of construction work includes: 'the installation, commissioning, maintenance, repair or removal or mechanical, electrical, gas, compressed air, hydraulic, telecommunications, computer of similar services which are normally fixed within or to a structure'.

In turn the definition of a 'structure' includes: 'any fixed plant in respect of work which is installation, commissioning, decommissioning or dismantling and where any such work involves a risk of a person falling more than 2 metres'.

If we accept that the parliamentary draughtsmen intentionally left out 'maintenance' from the latter definition, this implies that maintenance of fixed plant is not subject to the Regulations. However, most installation, commissioning, decommissioning or dismantling of the plant would be covered by the Regulations. To avoid confusion in the future, the industry needs to be clear as to what is the correct interpretation. I increasingly find examples of organizations attempting to apply the CDM Regulations where it is not necessary, for example in the maintenance of fixed plant.

Few in the industrial sector have fully appreciated the significance of the above definitions and the implications they have in relation to plant and services. It seems right in principle because a person falling 10 m from a machine under commissioning will suffer the same fate as someone falling a similar distance from a roof under construction. Plant operators need to appreciate the difference under the CDM Regulations between work on fixed services and work on plant itself. The impact of this on the manufacturing industry in the next century will be significantly influenced by interpretations made now (i.e. what is 'condamable'?)

The solution

In summary, I see the solution to the perceived CDM Regulations problem as straightforward and only made complicated by ignorance and intransigence. Following the simple rules below will bring real benefits to the projects:

(a) understand more clearly what is 'condamable'
(b) integrate health and safety into the project
(c) place emphasis on pre-planning
(d) recognize that the usefulness of a plan is inversely proportional to its size
(e) see the Planning Supervisor as a team effort
(f) designers need a simple risk assessment technique
(g) avoid cowboy organizations
(h) see the difference between competence and resource—avoid the paperchase.

The HSE and all those with a serious interest in health and safety should be greatly encouraged by the considerable progress made in health and safety in the construction industry. Many organizations, particularly 'blue-chip' ones, recognize that application of the CDM Regulations is good for their business. Our ability as an industry to shake off the irritations at the edge of the CDM Regulations will leave core requirements which no one can argue with. Much will be solved by the HSE shaking off its understandable inhibitions and continuing to answer the ongoing questions which cause such concern and misconception. In doing so, it will satisfy the curious and reassure the nervous. The rewards for the industry are enormous but within reach. Let us all aim to share in that.

Appendix 1: examples of designer risk assessment

Activity, design, process, material: Describe, including drawings, specifications, clauses, references, etc.	**Hazard/risk:** Outline the potential to cause harm as a result of column 1	**Initial risk rating:** See matrix and information below	**Actions by designer to reduce risk rating:** (a) Is the initial risk rating > 1? If yes, consider as a priority the need to avoid and reduce risks—see (b) and (c) below (b) Can the risk be reasonably avoided by changing the design or specification? If yes, explain how and change it (c) Can the risk be reasonably reduced by changing the design or specification? If yes, explain how and change it (d) If the final risk rating is > 1, explain why and what others need to do to minimize the risk	**Final risk rating:** See matrix. If > 1 information required for plan/file	**Is the final risk rating 2 or 3?** If so, in what form has information been passed to the Planning Supervisor?
High density 7·0 N/mm^2 masonry blocks, 215 × 440 × 215 mm	Each block weighs 40 kg and as such may cause a manual handling problem	3	*Option 1* (a) Yes—will effect a number of bricklayers (b) Risks can't be avoided. There is a need to maintain 215 mm load-bearing capacity and a fair-face finish (c) Yes—medium density 7·0 N/mm^2 solid blocks of 100 mm thickness can be used to form two skins (215 mm) with flexible bed joint reinforcement. The weight of each block is now 13·7 kg, a reduction in weight of > 26 kg per block *Option 2* Use lightweight blocks	1	No
Cleaning of roof lights	Access	3	(a) Yes (b) No (c) Yes. Access should be achieved via a mobile platform or scaffolding positioned next to the valley of the roof (d) Access details to be entered in file to alert maintenance workers	2	Yes

Severity \ Likelihood	H	M	L
H	3	3	2
M	3	2	1
L	2	1	1

Severity	*Likelihood*	*Risk Rating*
H—Fatality, major injury causing long-term disability	H—Certain or near certain	3 = High risk—action required
M—Injury or illness causing short-term disability	M—Reasonably likely	2 = Medium risk—action required unless good reason
L—Other injury or illness	L—Very seldom or never	1 = Low risk—no action required

4. Design for safety

Michael A. Williams, *Health and Safety Consultant*

Introduction

Traditionally, the brief for consulting engineers, architects and others has been to concentrate on meeting the needs of clients by producing completed structures, buildings or mechanisms that fulfil their intended functions, such as containing or supporting people, materials, plant, other structures, highways and so on. In most cases it is likely that the emphasis would have be on the 'completion' aspect, so that actual construction methods and safety may not have figured prominently in the thoughts or concerns of designers. Decisions on how site work would be carried out were considered the sole prerogative of contractors and site supervision, and checking would be focused on ensuring compliance with the design, specifications and dimensions. Perhaps this was the natural outcome of the way that the construction industry had developed over the years, with consultants designing and contractors building, each party confining themselves to what was considered to be their area of expertise and without any direct contractual link between them. Having won their work by competitive tendering based on cost, contractors then built the projects under contract to the clients and carried full responsibility for all aspects of the work, including health and safety, up to the point of completion and handover. As well as these contractual duties, a wide range of Acts and Regulations put the responsibility for construction safety directly on contractors, self-employed persons, site workers and suppliers of equipment, with no mention or involvement of the consultants, architects, surveyors or designers who drew up the specifications, bills of quantities and designs which were to be used in the construction.

However, the traditional roles of the parties to construction work and the separation of their functions and responsibilities has been clearly identified as a root cause of many site accidents, collapses and failures. These have often caused serious or fatal injuries, but, if only involving loss of time and replacement of materials, they created considerable costs and other problems. The construction industry has not helped by perpetuating the demarcation between consultant and contractor, and as a result some designers have lacked sufficient direct involvement and familiarity with construction techniques, materials and structures. Such knowledge would have helped them in their work and led either to anticipating construction safety problems or to eliminating them, a process which is now required by law. The important point which emerges is that the need for designs which are safe to build and maintain, not just on completion, has become an overriding priority reinforced by the arrival of the Construction (Design and Management) (CDM) Regulations 1994 (Health and Safety Executive, 1995), in which specific duties to ensure safe construction were laid on designers as a named group for the first time.

Duties of designers

Before going on to consider designs which have safety as an intrinsic component, it is as well to look at what is meant by 'design' and 'designer'. The CDM Regulations have helped considerably by providing definitions of both those terms, and 'design'—related to any structure—is held to include not only drawings and design details, but also specifications and bills of quantities. This is a very important extension of what has generally been understood to constitute a design and now brings into scope all materials, substances, equipment or articles specified in detail for use in projects, with all the safety implications that are involved.

When examining the definition of 'designer' set out in the CDM Regulations, it may not be clear on first sight who is held responsible for a 'design' as every person in an organization or consultancy who produces drawings, calculations or specifications may seem directly liable. Closer study shows that the 'designer' directly responsible for compliance is whoever—whether an individual or company—is carrying out the design work as a trade or business and has been formally appointed under the CDM Regulations. This generally means that a firm of consultants or architects will be the 'designer'. However, those appointed, if not in business as individual designers, have the duty of ensuring that all their employees and everyone else under their control who is doing design work complies with the requirements of the Regulations. This means that even though employees in a design consultancy are not personally liable for breaches of the CDM Regulations, their employers could be held responsible for their actions or omissions if safety during construction is proved to have been affected by some aspect of the design.

Remember also that the broad interpretation which the Regulations give to the term 'designer' includes not only the architects and engineers involved with the main project, but also plant and equipment designers, surveyors specifying articles or substances, contractors involved with 'design and build' work, falsework and shuttering designers, landscape architects and any in-house designers for subcontractors.

Having looked at the background and the definition of 'designer' and 'design', it is now time to look at what is currently expected and required of designers by law. As mentioned above, many designers have already been working on the lines of the CDM Regulations by virtue of their training and experience, and an understanding of the realities of construction work, so that they will have a clear picture of the way their scheme is likely to be built, including the assembly of individual components under site conditions.

In setting out the duties of designers, the CDM Regulations are in effect codifying best practice so that accidents and losses of all sorts are prevented and designers have to produce what is reasonably practicable and reasonable in the light of current knowledge at the time the design work is carried out. This means that, while on the one hand they would not be judged on the basis of later improvements and developments in techniques and materials, on the other hand there is a firm expectation of gradual improvements in designs and safety standards with time. Before leaving the concept of 'reasonable practicability', it is as well to establish some of the key elements that are required of designs:

- ☐ comply with current law
- ☐ follow appropriate industry standards
- ☐ checking by second, preferably independent, parties
- ☐ 'constructability' (to be discussed in more detail below)
- ☐ provide information on special features or identified risks
- ☐ use of standard solutions based on *safe* practice applicable to the particular project. (The emphasis on 'safe' may seem unnecessary but too many failures and accidents have resulted from slavishly copying previous designs which were either inappropriate for a later scheme or had insufficient safety margin for differing site conditions, loadings etc.)

The duties of a designer as set out in Regulation 13 of the CDM Regulations are:

(a) design to avoid foreseeable risks to the health and safety of anyone doing construction or cleaning work on a structure, or any person affected by their work
(b) reduce and control risks at source if avoidance is not possible
(c) give priority to measures which will protect all involved on the site, rather than individuals
(d) make sure that adequate information is included in the design about any residual risks, that is, anything that could create health and safety problems after all that is reasonably practicable has been done under (a), (b) and (c)
(e) co-operate with other designers and the planning supervisor working on the same scheme to ensure joint compliance with the CDM Regulations and other legal requirements.

There are some occasions when the CDM Regulations do not apply but it is essential to understand that the requirements of Regulation 13 still fall on designers, for example when work is being done for domestic clients, schemes are of very short duration or they involve less than five persons. In those cases, planning supervisors and principal contractors do not have to be appointed and various other requirements are waived, but, very importantly, designers still continue to attract the full force of the Regulations. In the case of demolition work, the CDM Regulations always apply and any design work involved is of course subject to them. This argues that to produce schemes for any construction work, designers must possess adequate experience and knowledge of the precise field in which they are working, which can be summed up as 'competence' and which the CDM Regulations require. It is not just the theoretical aspects that are essential but also an understanding of the way in which their scheme will be built, the construction techniques that will be used, and the interaction between their design and local environment. For example, in the case of structural steel frames, familiarity with detailing, fabrication methods, connections, site handling and erection methods will be essential to ensure a sensible understanding of the safety issues and hence to be able to comply with the requirements of the CDM Regulations set out above. In this context, an individual designer who has been concerned mainly with highway or coastal defence works may not possess the required degree of competence to work next on a complex high-technology building.

It goes without saying that knowledge of relevant safety legislation is an essential part of a designer's resources, as is the ability to carry out hazard identifications and risk assessments to determine the extent of any problems which might arise, together with knowledge of relevant mitigation and control measures.

Designing for safety

It is because of the unique contribution which designers make to projects that they are best placed to ensure that health and safety is an integral part of the work from start to finish, but to be successful, they must accept the role which the CDM Regulations now require and accept that competency comes only with a comprehensive practical as well as theoretical knowledge of the field in which they operate. To incorporate all of this into their work designers now have to take a broad view, currently summed up very well by the term 'constructability', which means that a structure is not only capable of being built and able to fulfil its end purpose, but that the whole process can be done with optimum safety.

First must come an ability to recognize hazards which may be caused by construction of their designs, where 'hazards' are understood by the generally accepted defnition of 'something with a potential to cause harm', either to those directly involved in building their design or anyone else who might be affected in the vicinity, as required by the CDM Regulations. This process should be aided by information about the site which the same Regulations demand should be provided by clients, and by holding 'constructability reviews' at early stages in the planning and design of all schemes. Having outlined the designs, discussions should be held at an early stage between all those involved to consider where and how structures will be built and to evaluate the safety implications of design proposals on the site, its surrounding environment, the eventual workforce and, where appropriate, the interaction between separate elements of the structure and the project as a whole.

In the light of these discussions, various actions might be taken, ranging from acceptance; slight modifications to improve safety; alterations and additions with safety information on residual risks; or, in an extreme case, major changes or cancellation to remove unacceptable levels of risk. It may be that, after risk assessments have been done, particular safety problems have been identified—say, in the the stability of high-technology structures during their erection or the use of newly developed materials. If it is felt, after careful evaluation on the basis of 'reasonable practicability', that to proceed with the design is fully justifiable, a specific construction method with safeguards should be adopted to keep risks down to an acceptable level and then indicated to potential contractors at the pre-tender stage as the preferred procedure.

This is not to say that as a general rule designers are now expected to impose construction methods on contractors or to usurp their role in making decisions on how to carry out site work. What is essential is that constructability reviews should be conducted and risk assessments made on the basis of accepted construction techniques, and only when special methods or precautions are needed should a particular method be indicated and included in contract documents. This would still leave it open to a contractor to propose an alternative method for consideration and agreement.

In other cases, where it is reasonable to expect conventional methods of building, then if problems still remain after foreseeable risks have been avoided or dealt with at source on a 'reasonably practicable' basis, then the designer's duties will be met by providing suitable information to ensure safety, as required by Regulation 13(2)(b) of the CDM Regulations.

Constructability and maintainability

Having emphasized the importance of 'constructability' as an essential part of design work, it may be helpful to look at a simple example which incorporates it into the process of 'designing for safety' and illustrates the way designers must now consider construction work. A situation might arise where the design proposal for a structure calls for foundations consisting of augured piles and reinforced concrete pile caps. Information from site investigations is obtained which shows that the foundations will be built in ground which has pockets of heavy contamination, at which stage the original design should be re-examined with great care. An alternative method might then be put forward to use driven piles so as to displace the material rather than excavating it, thereby not only avoiding exposure of the workforce and others to the contaminated soil but also removing the problems of loading, transporting and disposal of toxic material. As well as the obvious safety advantages, the combination of hazard identification and mitigation measures could have important cost and efficiency benefits. Taking this example further, recognizing that pile caps below ground level would involve the excavation and support of contaminated material, serious consideration should be given to constructing them above ground level, again with safety, cost and time benefits.

As well as the attention designers must give to the construction phase, what might be termed 'maintainability' is an additional duty, as a look at the definitions given in the CDM Regulations show that an extremely wide range of activities are classed as 'construction work'. Designers must therefore be aware of the inclusion not only of the expected building, alteration, repair and redecoration work, but also of cleaning, maintenance, dismantling and demolition. As regards the 'cleaning' of certain parts of structures such as glass walls, windows, ceilings and roofs, special consideration must be given to this by designers as, importantly, their duties now include making provision for the safety of persons involved in this work on completed buildings where there is a risk of a fall of more than 2 m. By extension, the use of fragile materials for structural elements of buildings, such as roof lights, is now seen as unacceptable unless prevention of falls and general safety during access for cleaning or maintenance can be dealt with in other ways.

In the past there has been a tendency for some structural designers and architects to overlook this aspect so that on completion and occupation, considerable practical and safety problems are encountered with routine cleaning and maintenance work. In some cases, which regrettably have included major bridges and prestigious buildings, this has resulted in having to install specialized cleaning or access equipment at a later stage, often with great difficulty and at disproportionate cost, on structures which were not originally designed to accept the additional loads and whose aesthetic appeal has then been permanently impaired. If a large building is being designed which has a major feature such as a glazed dome or barrel roof over a multi-storey height atrium, then careful thought must be given not only to how it will be built but also how the routine cleaning and redecoration can be done safely, both internally and externally. A considered and well-engineered solution should provide safe access for construction as well the maintenance of the large areas of glazing. Any provisions for cleaning work would also have to ensure safety for the building's occupants and users, because if the whole of the atrium has to be closed for safety reasons throughout each cleaning session, then serious financial loss might be caused if, for example, shops or offices depending on the atrium for access had to be closed.

Hazard identification and assessment by designers

Some of the types of site work which should be considered by designers when setting out to identify hazards and their likely effects during construction of their designs are as follows:

- site planning—site access and the surrounding environment
- site clearance, minor demolition, site investigation and survey
- excavations and foundations
- conventional structures in concrete, steel, masonry, timber, etc.
- special structures, for example air-rights buildings, tension-cable-supported buildings, complex prestressed concrete structures, chimneys, cooling towers
- highways, railways, bridges, river and sea defence works, cable- and pipe-laying
- cladding, roofing, windows, curtain walling and large glazed areas of walls and roofs
- refurbishment, alteration and extension of existing buildings and other structures
- major demolition projects.

A great deal of very useful information is given in the Construction Industry Research and Information Association (CIRIA) Report No. 166 (CIRIA, 1997), which sensibly introduces the subject matter with the comment 'Designers are intended to use this handbook as a first point of reference. They should then apply their own thinking to particular projects in order to identify the hazards ...'. This is eminently sensible and effectively endorses the approach advocated earlier that it is the designer's own responsibility to design for safety on the basis of appropriate experience.

As part of the preparation that should now be carried out during planning and outline design for a project, the first stage should be to obtain as much information as possible about the site and its surroundings, most of which should be available from the pre-tender Health and Safety Plan prepared by the Planning Supervisor from information supplied by the client or the client's agents. Eventually, the Health and Safety Files required by the CDM Regulations for completed structures will become a useful source of information when new works are to be carried out. However, they are not likely to be available as useful source material for some years yet, until a large enough number of post-CDM structures have been completed. In the meantime, there can be no substitute for the 'reasonable enquiries' about sites and premises which the Regulations require clients to make to obtain and provide information. Enlarging on the items listed above, the following areas for investigation and gathering of detailed information are suggested.

Site surroundings and local environment

Access road limitations might include difficulties due to an approach along narrow or load-limited sections; turn-offs from a dual carriageway; heavy peak flows and congestion; small radius turns unsuitable for long loads such as bridge beams; height restrictions; access to be maintained for adjacent businesses, schools, hospitals or the emergency services. Railways, motorways and rivers also pose problems, the two former imposing limitations for crossings and proximity to work and plant positions, the latter sometimes causing flooding, high groundwater levels or difficulties over maintaining a navigation. On very constricted urban sites, there could well be problems over noise, dust, night work or where cranes can be sited so as to avoid infringement of 'air rights', such as when tower cranes might slew their jibs or counterweights over adjacent private property. When working on refurbishment projects the safety and security of residents, particularly children, must be taken into account, as must their need for access and freedom from excess noise levels, pollutants, etc.

Site clearance, demolition and investigations

The contents of a site can often pose serious problems, such as toxic, pathogenic, flammable or explosive materials in buildings, tanks or the ground itself; live underground and overhead services; existing foundations, cellars or basements; unstable buildings on or adjacent to the site; drains, culverts and watercourses subject to surcharge or flooding. For demolition work, many of the same hazards are also present, to which can be added the removal and disposal of materials such as asbestos, lead paints, chemical and other substances. Although preliminary surface surveys are not generally classed as CDM Regulations work, ground investigations are, and any trial pits, test bores, etc., should be assessed carefully, with appropriate safeguards.

Excavations and foundations

As for site clearance (see above), the contents of the ground at a site must be determined with certainty before work begins and any services and

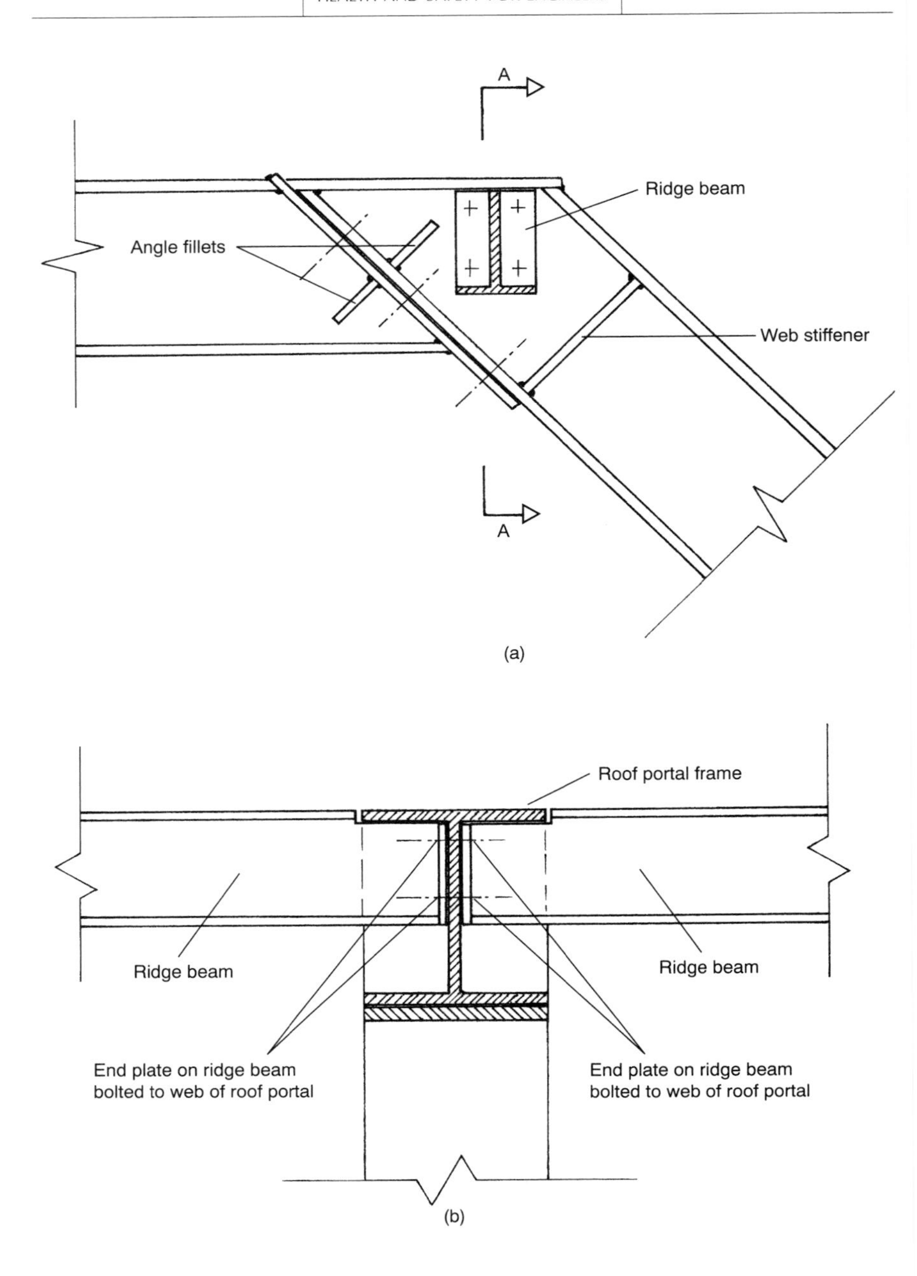

Fig. 1. Original roof structure showing the connection between the ridge beams and portal: (a) elevation; (b) section A–A

contaminants located and marked, with suitable information made available. Safe systems for excavation support will be necessary, taking into account neighbouring structures, underground services, drainage and bearing capacity for plant. Once large excavations, pits, trenches, etc., have been established, edge protection and safe traffic routes for persons and plant must be established and measures adopted to prevent falls. Any trenches, underground chambers, basements, etc., will have to be assessed for confined space working, and possible flooding from nearby watercourses or sewers prevented. A further task will be the safe disposal of pumpings from contaminated sites and the disposal of toxic material to licensed dumps in suitable transport.

Erection of conventional concrete structures

For straightforward mass, reinforced or simple prestressed concrete buildings and other structures, the main points of concern are likely to include the stability and adequacy of temporary works (formwork, falsework, shoring and other supports); use of transport, plant and equipment; prevention of falls; and noise, dust, irritants and chemicals. Where pre-cast elements are to be assembled, the use of cranes and lifting equipment, often on very constricted sites, creates risk. Design should take into account the extent to which work at heights for in situ construction might be mitigated by use of pre-casting; design of connections to limit work in dangerous positions with large crane-supported loads to position and manipulate; care in the positioning of items such as columns to allow adequate room for shuttering; elimination of work in unsupported excavations behind shutters for retaining walls; reinforcement cages designed for prefabrication and lifting into excavations; provision of safe lifting attachments on pre-cast items, and marking of mass and centre of gravity to aid safe slinging and positioning; and safe access for stressing and grouting operations.

Steel erection

For steel erection, the main problems are still working at heights, instability of incomplete structures, and the storing, stacking and moving of components. These are key areas for 'constructability' assessments to prevent risk and designers can achieve very considerable improvements by: provision of information on the need for temporary bracing before full stability is assured; arranging for prefabrication of subassemblies at ground level to reduce work at heights and crane time; and provision at an early stage of edge protection, flooring and stairways to prevent falls and to provide 'collective safety' for other trades. There is also still a need for designers to have a better understanding of the problems of making connections on structural steel frames during their construction.

Special structures

It is in this field that the greatest demands are made and most reliance placed on the abilities of designers to consider 'constructability'. For innovative structures, great care has to be exercised to envisage the whole of the construction process and, after avoiding risk wherever possible, to incorporate safety procedures and information into the design. Main problems likely to be met include: temporary structural instability; overload of components; uncertainty over the sequence of building; difficulty of access to connections; inadequate foundations and bearings; errors in or substitution for specified materials; environmental effects; and storing, handling and lifting materials and components.

Highways, railways and bridges

Additional difficulties found with this type of work, over and above the project or structure itself, are caused by the effect of the surrounding environment and vice versa. For highways and bridges, proximity to live traffic and the problems of maintaining routes and flow levels have to be taken into consideration at the design stage except, perhaps, not where it is completely new construction on a separate site. Even then, road access to and from the site, haul routes and transport of large components must be planned in advance for feasibility. Bridges over existing roads, rail or rivers bring their special problems of construction methods and sequence, together with avoidance of falls of persons or materials, separation and security of the site and the adjacent undertaking. As an example of how safety can be improved by design and planning, many bridges are now being built by launching, sliding or rolling in on transporters, which offer many advantages, including much of the structure then being erected at ground level and off-line, avoiding work near or over live roads, railways, rivers, etc. For railways, many tasks which on standard sites would create no special difficulty will present serious problems, including: the lateral and vertical clearance from rail traffic; danger from live rails and overhead conductors, taking particular account of arcing distances; positioning of plant, particularly cranes, to avoid danger in case of overturning or jib failure; timing and cost of possessions for essential encroachment on or over tracks; and storage and supply of materials to the work position. River and sea work will require specific details for: water levels and flows, flooding and tides; safe access; preventive measures against falling and drowning, including protective equipment, safety boats, etc.; liaison with the emergency services; communications; and provision for shift and tidal work.

Cladding and roofing

For this work, probably more than elsewhere, the designer can effect very marked safety improvements in terms of both 'constructability' and 'maintainability'. As large areas of cladding and roofing, particularly transparent or translucent windows, walls or roofs, require regular attention after completion, the initial design and choice of material will have a direct bearing on safety during building and in use. Considerations will include: specification of non-hazardous materials; avoidance of fragile materials; arrangements for safe access and prevention of falls during erection; provision of permanent access equipment for maintenance and cleaning; adequacy of anchorages and fixings for access devices; safe positioning of fixings for wall and roof elements for ease of installation; arrangements for removal and replacement of damaged sections; permanent edge protection on roofs; and planned procedures for removal and replacement of plant in or on roofs, for example, air conditioning.

Refurbishment, alterations and extensions

Schemes involving existing premises and structures require exemplary information searches and verification of conditions on which designs for safe work can be based. Existing 'as-built' record drawings and specifications can rarely be relied on, and serious changes in strength and stability may have taken place. Detailed inspections and surveys will usually be necessary to establish the present condition of buildings and their contents. Problems include: weakening of the structure due to damage, deterioration or alterations; shortcomings of original construction, such as omission or misplacement of reinforcement; presence of toxic, flammable or explosive substances; live services and drainage; limitations of access; party wall and adjacent building vulnerability; confined spaces; foundation, soil and groundwater problems; access, storage and working space limitations; and restrictions on noise, dust and working hours in occupied premises.

Case studies and examples of design-induced failures and problems

Steel-framed building accident

Steel erection is one of the construction sectors most subject to problems directly attributable to designs which, although not technically at fault in that the completed structures would have performed successfully, have created hazards resulting in premature collapse or accidents, some fatal. For example, cases are still seen of holding-down bolts being slackened to tilt columns out of plumb, with all the risks that are then likely for partial or total collapse, to insert beams between the flanges of columns. This arises when the design calls for connection of beams to the webs of columns where the overall length of the beam is greater than the distance between the flanges. An error of this type is the mark of a designer who lacks competence and practical experience, and is the more regrettable because simple solutions usually exist which are not only safer but often save time and effort.

In the example shown (see Fig. 1), the original roof structure design consisted of medium-size portals linked by eaves beams. Because of transport, handling and access problems, the portal frames were designed to be fabricated in three parts for site jointing, the joints strengthened by web stiffeners and angle fillets. The eaves beams were then to be joined to the portals by bolting their end plates inside the boxes formed by the stiffeners and fillets, as shown. Unfortunately, this then created the problem referred to above, as it was not physically possible to position and bolt up the end of the eaves beam unless one of the portal frames was tilted away on its base. While this operation was being attempted, control of the frame was lost, the tilted portal fell and an erector working at the joint position fell with the steelwork and was killed.

A simple and well-known connection detail (see Fig. 2) could have been designed, using a slightly shorter beam which would have had the same engineering performance as the original. This has the added advantage that no tilting is needed and the shelf bracket allows the beam ends to be landed and supported, reducing risk for erectors, who can then concentrate on bolting up the connections without having to manhandle steelwork or be placed in danger due to temporary instability.

Pipeline problems

Raw water was to be conveyed several kilometres overland from a new river intake to an existing treatment works and the water company designer decided that a steel pipeline of 800 mm diameter, each pipe 13·5 m long to reduce the number of joints, would provide the quantities needed. The design also provided for 600 mm diameter branch 'T' pieces every 270 m for access. Due to the aggressive water, the pipes were bitumen-lined and, to expedite matters, the water company ordered, purchased and supplied the pipes and other materials, such as a chemical primer for application to joints, to a contractor who was employed only to excavate the pipe trench to an average depth of 4 m, then lay, joint and backfill over the completed pipeline. During construction, various hazards were identified which had almost resulted in a serious accident and were a direct consequence of the water company design. Enforcement action was taken, the work was halted and changes to the specification and working arrangements were introduced to remove hazards and to provide safeguards before construction was allowed to continue.

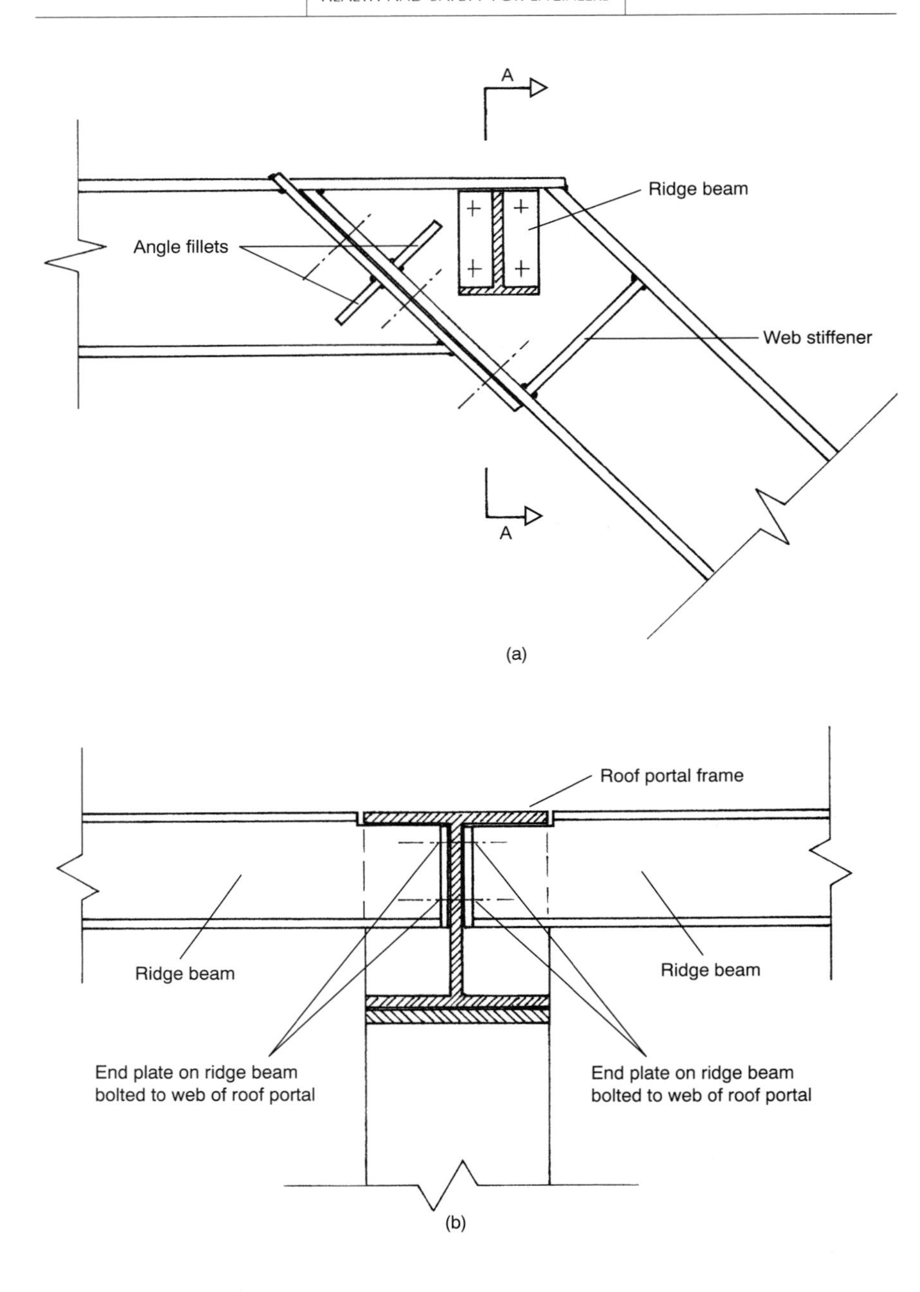

Fig. 1. Original roof structure showing the connection between the ridge beams and portal: (a) elevation; (b) section A–A

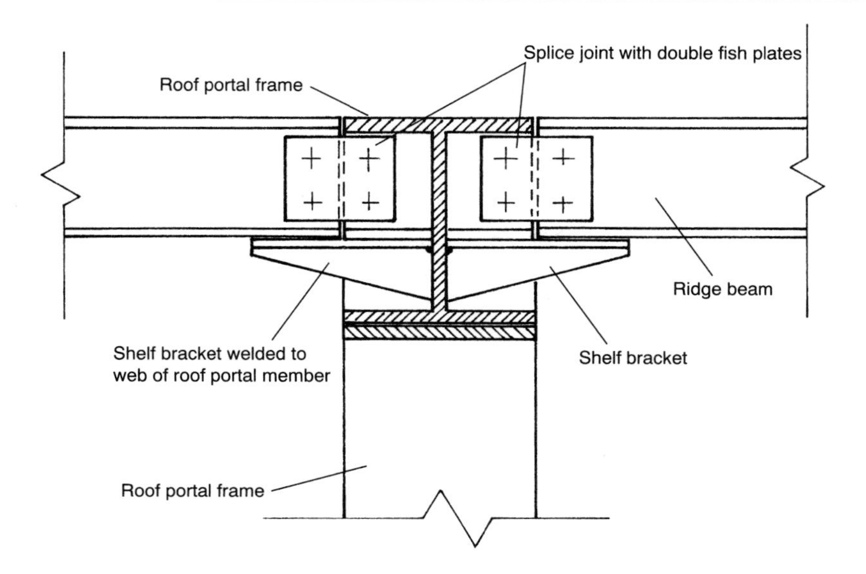

Fig. 2. Alternative roof structure using shelf brackets and fish plates: detail of section A–A

Problems included:

- The pipeline diameter was less than the 900 mm minimum accepted for entry by persons.
- The design called for both external and internal welding of the socket and spigot joints, requiring entry for welding in a confined space (minimum distance to joint in pipe 13·5 m).
- Welding mild steel by electric arc created dangerous fume concentrations.
- 'Fettling' of the internal weld with angle grinders was required to remove slag and smooth surfaces. This not only generated dust and fume but was also an ignition hazard.
- After fettling, a primer supplied by the client was to be brush applied. The primer contained over 70% volatiles, including xylene and trichlorethylene, both hazardous substances when used in a confined space, causing toxic and explosive dangers. No hazard information for the primer was supplied to the contractor.
- The bitumen lining was specified to be made good over internal primed welded joints by hand trowelling of molten bitumen. The bitumen had to be heated at the trench side and conveyed through the 600 mm 'T' pieces and then along the 800 mm internal diameter pipeline to joint positions, any heat and plasticity loss being made good with liquified petroleum gas (LPG) burners inside the pipe.

There were several hazards, some cumulative, caused by the failure of the designer to understand the consequences of the construction processes and materials that were imposed on the contractor. A prior appraisal of the likely hazards during construction should have resulted in a change of design to one that eliminated the need for internal entry and the use of potentially dangerous materials and methods.

Steel box-girder bridge collapse during construction

Although it took place in 1969, the tragic collapse of the Cleddau Bridge at Milford Haven, south Wales, during its construction is still relevant and highlights the dangers of structural design concentrating on the performance of completed structures to the neglect of safety during the erection phase. At Milford Haven, the massive steel box-girder bridge was being built incrementally outwards from one of the high-level piers when the box structure failed over the pier support. The report by the Committee of Inquiry into the Basis of Design and Method of Erection of Steel Box-Girder Bridges (the Merrison Committee) (1973) concluded unequivocally that the primary cause of collapse was that 'the diaphragm as designed was simply not strong enough to resist the forces to which was subjected' while the next span was being cantilevered out and that this weakness had not been revealed during design checks. It also emerged that some of the steel in the box-girder was of a lower grade than specified.

Floor loadings on multi-storey concrete-framed buildings

Another example of the need for designers to understand how structures are built and to provide information on maximum sustainable loads during construction can be seen in the case of reinforced concrete floors of multi-storey buildings designed for very low in-service loadings, such as open-plan offices. Construction of the first floor using falsework and shuttering supported off a solid ground floor slab will usually present few problems. However, for the second and subsequent floors, there is clearly a strong likelihood of excessive, probably point, loads being imposed by falsework propping systems supported off lower floors which have not reached full strength and have been designed for low-level distributed loadings. To avoid the possibility of overload and possible collapse, the designer has a duty to anticipate the problem and to ensure that sufficient accurate information on any limitations or risks is clearly expressed and conveyed in contract documents, health and safety plans and drawings.

The effects of the CDM Regulations on design

From the foregoing discussion of designing for safety, the following key points emerge:

- ☐ Failures, losses and accidents are caused if designs do not take account of the realities and effects of construction work rather than concentrating solely on satisfactory performance after completion, a process which has been termed 'constructability'.
- ☐ Not to consider 'maintainability' in designs is a source of additional later costs and causes serious problems in carrying out maintenance, cleaning and repairs.
- ☐ The CDM Regulations do not impose harsh restrictions on designers—they encourage and confirm good practice, while preventing low-grade work. This generally means that cost cutting and unfair competition at the expense of safety should be prevented.
- ☐ The Regulations rightly emphasize the need for designers to abandon the 'ivory tower' attitude by becoming competent in construction techniques in their field of activity so that their designs will be safe and effective during building, not just on completion.

Critical comments have been made on the effects of the CDM Regulations, among them the proposition that they could stifle innovative structural and architectural designs. It is hoped that by explaining and illustrating the duty of designers to consider safety at all stages of construction it will be seen that to follow the procedures of the CDM Regulations and adopt the underlying philosophy is not a limitation on progress and should not create problems. As design concepts move forward and new materials and techniques are introduced, there should be no obstacles if the basic theme of 'safety during and after construction' is incorporated into designs. After all, the construction and allied industries have been constantly changing and adapting to keep pace with technology, and the greater emphasis on integral safety which the Regulations require is no more than a sensible process of consolidating good practice and requiring higher, rather than lower, standards of construction across the industry while at the same time reducing and controlling losses of all types, particularly accidents.

Bibliography

Committee of Inquiry into the Basis of Design and Method of Erection of Steel Box-girder Bridges (1973). *Report of the Committee*. HMSO, London.

Construction Industry Research and Information Association (1997). *CDM Regulations—Work Sector Guidance for Designers*. CIRIA, London, Report 166.

Health and Safety Commission (1995). *Managing Construction for Health and Safety. Construction (Design and Management) Regulations 1994. Approved Code of Practice*. HSE, Sudbury, L54.

5. Risk assessment

John Green, *Cambridgeshire County Council, Cambridge*

Introduction

As an engineer, going about your work, you come into daily contact with others of different disciplines going about theirs. It is this combination of working together in any number of situations that can combine to create risks to health and safety, and it is these risks that legislation seeks to overcome.

Why are formal risk assessments needed?

Before 1992 several health and safety regulations existed that required risk assessments to be undertaken. By the introduction of the Management of Health and Safety at Work (MHSW) Regulations 1992 the law required further assessments to made, and it is principally these regulations that require us to carry out risk assessments. Hazards must be identified, risks assessed and practicable procedures produced, so that anyone who might be affected by the work we do is not put at risk. 'Anyone' means fellow employees, the clients served, the consultants engaged, the contractors employed and the public in general, whether visitors or people affected by the work done. Risk assessments are required to take all of this into account, and where premises are shared employers might need to co-operate to provide adequate assessments.

The objectives of risk assessment are as follows:

- to reduce accident frequency and severity
- to minimize damage to property/facilities
- to provide safe and healthy working conditions
- to comply with legislation.

The benefits to be derived from risk assessments are:

- the elimination of pain and suffering
- to identify training needs
- to formulate safety plans
- to prioritize allocation of resources
- to provide documentary evidence
- to reduce financial losses from operational disruption, reinvestment in recruitment and training, fines in criminal proceedings, settlements in civil proceedings and increased insurance premiums.

Definitions

Risk assessment Risk assessment determines systematically what the hazards are, the probability of harm occurring, and the possible consequences of that harm and its severity.

Hazard A hazard is the potential for harm. Vehicles travelling on a road are a hazard, as is coffee spilled on a tiled office floor.

Risk A risk is the chance or likelihood that someone will be harmed by the hazard (Fig. 1). There is a risk when driving on the road, or walking across a wet floor. Thus in simple terms risk assessments determine the hazards, the probability of harm occurring and the possible consequences. This, in turn, allows controls to be identified, and they can then be introduced to reduce the risk, or its effects, and thus provide the information for the production of safe systems of work.

The law

Risk assessment is not a new concept. Case law earlier this century established the fact that risks needed to be assessed. More recently, legislation introduced the concept in such regulations as the Control of Asbestos at Work Regulations (1988) and the Noise at Work Regulations (1989). However, the 1992 legislation introduced the concept of risk assessment. This completed the so-called 'Six pack', that is

- ☐ Management of Health and Safety at Work Regulations
- ☐ Workplace (Health, Safety and Welfare) Regulations
- ☐ Provision and Use of Work Equipment Regulations
- ☐ Personal Protective Equipment at Work Regulations
- ☐ Manual Handling Operations Regulations
- ☐ Health and Safety (Display Screen Equipment) Regulations.

The most important of the six is the MHSW Regulations which clearly set out the principle of risk assessment. The Regulations are accompanied by an Approved Code of Practice which provides practical guidance on their application.

It should be noted that not complying with the Code is not an offence, but if it has not been followed, a court will need to be satisfied that it has been complied with in a better way. Previous legislation in the form of the

$$\text{Risk} = \begin{matrix}\text{Hazard}\\ \text{effect}\\ \text{(severity)}\end{matrix} \times \begin{matrix}\text{Likelihood}\\ \text{of}\\ \text{occurrence}\end{matrix}$$

Fig. 1. Definition of risk

Health and Safety at Work etc. Act 1974 (HSWA), although it did not use the phrase risk assessment, implied the need for risk assessment in its Sections 2 and 3:

> *Section 2* It shall be the duty of every employer to ensure, so far as is reasonably practicable, the health, safety and welfare at work of all his employees.

> *Section 3* It shall be the duty of every employer to conduct his undertaking in such a way as to ensure, so far as is reasonably practicable, that persons not in his employment who may be affected thereby are not thereby exposed to risks to their health or safety.

'Reasonably practicable' requires the balancing of the cost in time, effort and resources, against the risk (Fig. 2). What is reasonably practicable will in due course vary as knowledge and awareness grows. The test of what is reasonably practicable is usually taken as how the average person would view the situation.

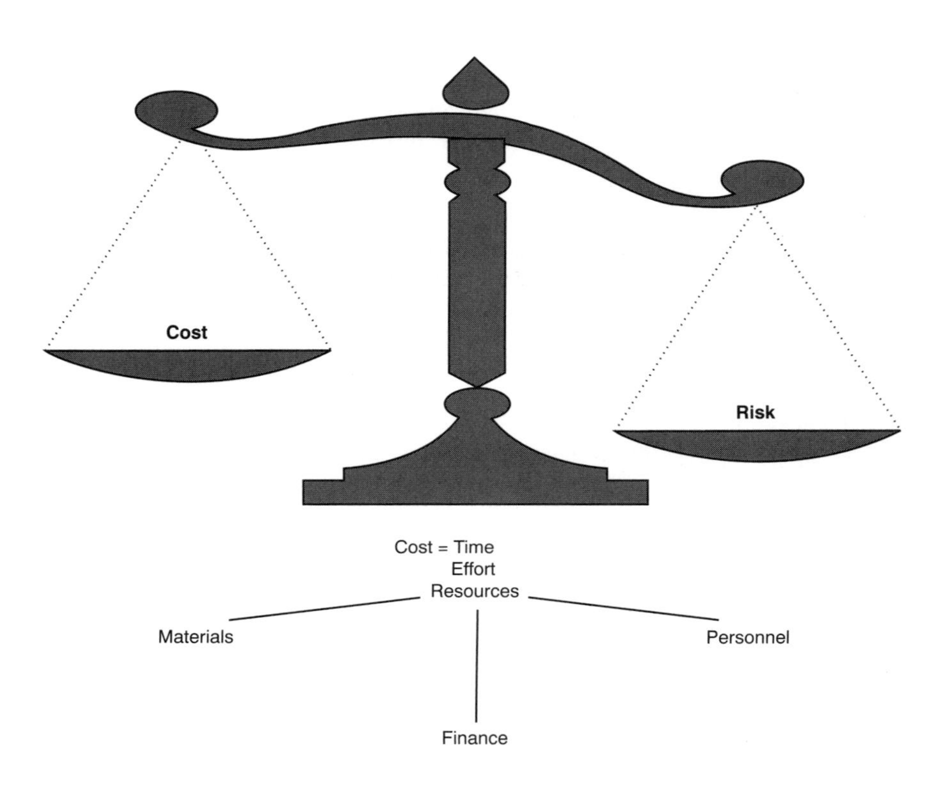

Fig. 2. 'Reasonably practicable' balances cost against risk

Regulation 3 of the MHSW Regulations sets out what is required with regard to risk assessment:

Section 1 Every employer shall make a suitable and sufficient assessment of:

(a) the risks to the health and safety of his employees to which they are exposed whilst they are at work; and
(b) the risks to the health and safety of persons not in his employment arising out of or in connection with the conduct by him of his undertaking for the purpose of identifying the measures he needs to take to comply with the requirements and prohibitions imposed upon him by or under relevant statutory provisions.

Section 2 places a similar requirement on self-employed persons.

Section 3 requires both employers and the self-employed to review their assessments if they are no longer valid, or significant changes have taken place.

Section 4 places a duty on employers who employ five or more employees to record the significant findings of their assessments and any group of their employees identified as being at risk.

The following are Regulations which contain particular requirements for risk assessment, and which are most likely to affect the majority of companies and institutions:

- Management of Health and Safety at Work Regulations 1992
- Manual Handling Operations Regulations 1992
- Personal Protective Equipment at Work Regulations 1992
- Health and Safety (Display Screen Equipment) Regulations 1992
- Noise at Work Regulations 1989
- Control of Substances Hazardous to Health Regulations 1994
- Control of Asbestos at Work Regulations 1987
- Control of Lead at Work Regulations 1980
- Construction (Design and Management) (CDM) Regulations 1994
- Construction (Health, Safety and Welfare) (CHSW) Regulations 1996.

The last two Regulations are, for obvious reasons, important to engineers. The previous chapter on the impact of the CDM Regulations sets out the key requirements, but both sets of Regulations impose strict duties as regards risk assessment on the various duty holders.

For the CDM Regulations the duty holders are:

- the Client or his agent
- the Planning Supervisor
- the Principal Contractor
- Other contractors.

For the CHSW Regulations the duty of assessing risk in the various areas covered is placed with:

- employees
- the self-employed
- people who control the way in which construction work is carried out.

There are also Regulations covering specialized risks such as major hazards, ionizing radiation, genetic manipulation and pregnancies. It should be noted that all health and safety law is criminal law and breaches carry the same penalties as other criminal law.

Perception of risk

Risks tend to be seen subjectively by the person considering them. While some people see a certain activity as a relatively low risk, given that appropriate precautions are taken, others would consider it highly dangerous whatever the precautions. You need to be aware that when individuals are asked to carry out assessments they may choose to ignore any of the relevant factors and merely make, using their own judgement, experience or bias, a bold statement that such and such an activity is a high or low risk. When considering risk assessment it is important to remember this fact and to provide a structured risk assessment system. In general, it is better to involve a group of people who work in the environment where the risks occur, to assist in the carrying out of the assessment. In this way individual bias is hopefully reduced and knowledge of what the actual risks are is given due consideration. Managers and supervisors are sometimes unaware of what is actually going on at the 'sharp end'.

It is also worth remembering that the term 'risk' tends to be viewed from the discipline in which the people considering the risk finds themselves, that is to say, financial, public service, politics, insurance, etc. From an engineering perspective, risk assessment is normally seen in terms of personal health and safety, risks to the environment, and to plant and equipment.

It is important to realize that the initial aim of risk assessment is to decide if the process, and/or hazard, can be eliminated altogether. For example, if a designer designs the steelwork of a building in such a way that it can be pre-assembled on the ground, then many of the risks associated with steel erection being carried out at heights can be avoided. If the hazard cannot be eliminated, then risk assessment provides the means whereby controls can be established and a safe system of work introduced.

The MHSW Regulations do not require new assessments to be carried out where they have already been completed to meet the requirements of other legislation, for example, the Noise at Work Regulations (1989).

Perception of risk requires the realization that it involves a common-sense approach to the hazards of the workplace and provides a means of identifying the risks resulting from those hazards. From this realization there needs to be a means of combating the identified risks in a practical and logical way.

Practical risk assessment

Whatever method is used to carry out risk assessments it should be seen as a means to an end, the elimination or the reduction of risks to an absolute minimum. Normally the complexity and perceived danger of the risks will decide the method to be adopted. While a simple high, medium or low system will suit many applications, it is obvious that certain industries by their very nature will require complex and rigorous risk assessment systems. If you are required to carry out an assessment the Health and Safety Executive (HSE) has produced a publication called *Five Steps to Risk Assessment*. The main points are given below. A flowchart is provided in Appendix 1. However, be aware of what is being undertaken. You will need to take a critical look at what can cause harm to people carrying out their work. You will need to decide if enough precautions have already been taken to prevent harm to the workforce, or whether more needs to be done.

Remember :

- ☐ *Hazard* means something that can cause harm—electricity, manual handling, plant or scaffolding.
- ☐ *Risk* means the likelihood that a person will be harmed—he or she will get an electric shock, will injure their back lifting a heavy weight, be knocked down by moving plant or, fall off an unguarded scaffold.

Once you have carried out your assessment, and assuming more is required to be done, you will need to look at how any existing procedures may require amending or modifying and/or new safe systems of work formulated and put into place.

Given below are the main points of the HSE's 'Five Steps'.

Step 1: look for the hazards

- ☐ Walk around the area subject to the assessment and look afresh at what may cause harm.
- ☐ Ignore the trivial and concentrate on significant hazards.
- ☐ Involve the people who work in the area and consult any representatives they may have. Ask them what they think the hazards are.
- ☐ Look at manufacturers' instructions and/or datasheets.

- ☐ Look at the accident book and talk to occupational health staff. This could help in establishing the likelihood and severity of any risks from incidents that have already taken place.

Step 2: decide who might be harmed and how

Think about everyone who is likely to be affected by the work going on in the area and who could be harmed by the activities, cleaners, visitors, contractors, maintenance people, the public and people in your workplace. Remember, Section 3 of HSWA places an obligation on employers to conduct their undertaking in such a way to ensure those who *might* be affected are not put at risk.

Step 3: evaluate the risks arising from the hazards and decide whether existing precautions are adequate or more should be done

Decide for each significant hazard, if it cannot be eliminated, whether the remaining risk is high, medium or low. Ask yourself the following questions:

- ☐ Have I done all the law requires? For example, guarded dangerous parts of machinery, provided safe access to and egress from a site.
- ☐ Are accepted industry standards being met? For example, declaring construction sites to be hard-hat areas.
- ☐ Is everything reasonably practicable being done to provide a safe workplace? That is to say balancing the risk against the cost in time, effort and resources.

If the work does not vary, and/or employees move from one site to another carrying out similar work, assess the risks from those hazards which can reasonably be foreseen, and provide common generic assessments. An example of a generic assessment is provided in Appendix 8.

If you share a workplace with others, ensure everyone knows about the risks your work involves and what precautions need to be taken. Conversely, they must also provide you with similar details of their risks and the precautions that need to be taken.

Remember, ensuring people are informed means keeping information simple and to the point, and in some cases information will need to be provided in languages other than English.

Step 4: record your findings

If five or more people are employed, assessments must be written down. Write down significant hazards and record your most important conclusions. For example, for electrical installations, insulation and earthing checked and found sound; or for fumes from welding, local exhaust ventilation provided and regularly checked. Tell your employees of your findings.

You are not required to show how you carried out your assessments, but you will need to prove that:

- ☐ A proper check was made.
- ☐ You consulted people who might be affected.
- ☐ You dealt with all the obvious significant hazards and took account of the number of people who could be involved.
- ☐ The precautions taken were reasonable and any remaining risks were low.

Assessments need to be suitable and sufficient. For example, full traffic management would need to be provided to carry out a repair on a motorway, whereas a small repair on a little-used country road would perhaps only require warning signs, and a set of cones, to be sufficient. The courts will ultimately judge what is suitable and sufficient, and what is not. Previous case history may be used when such judgments are made. Written documentation should be kept for future reference. It will help you if an HSE Inspector questions your precautions, or if you become involved in a civil liability case.

Remember, you can refer to existing documents which can provide information, health and safety manuals, policy statements and procedures. You do not have to repeat work that has already been done.

Step 5: review your assessments from time to time and revise them if necessary

However, assessments do not need to be amended for every trivial change, or even new jobs, but if new hazards are introduced, assessments must be carried out.

Workplaces contain many potential risks, but as long as regular maintenance is carried out, and hazards are reported and dealt with, there is little to be gained from conducting theoretical assessment exercises. So when should a fresh assessment be carried out? The following provide some of the reasons:

- ☐ There is new evidence or information that a risk could be controlled in a better way with little or no cost in time or trouble.
- ☐ As a result of a prosecution receiving publicity, it can be seen that current controls do not meet the legal requirements of the relevant legislation, due to either a change, or a new interpretation of the law.
- ☐ The passage of time has brought about change, new machines, new processes, substances, etc., and these have introduced new hazards.

- ☐ You are lagging behind in what is seen as accepted safe practice by others in your industry or same line of business.

Carrying out the risk assessment

How should you go about carrying out an assessment? Remember, there is no set format to follow. The law requires only that 'every employer shall make a suitable and sufficient assessment'; how it is achieved is left to the individual.

Simple risks can be assessed using a simple weighting and recording system, or you can use a computer-generated system, of which there are many on the market. For the more hazardous areas such as the petrochemical and nuclear industries, as would be expected, there will be far more formal and complex systems to follow.

Are you competent to carry out an assessment? You must be familiar with the risks for which the assessment is being undertaken. The HSE's definition of a 'competent person' is a someone who has 'the necessary experience, knowledge, training and instruction to do the job competently'. If you do not fulfil all of the requirements you should assemble a team of people who can not only carry out the assessments, but who can also formulate procedures and introduce the appropriate controls. Checklists are provided in Appendix 3.

Use of generic assessments

Generic assessments are those that cover a regularly repeated operation but with varying degrees of risk. In these cases one generic assessment can cover the whole of the particular area of work. An example of a generic assessment for carrying out emergency pothole repairs to a carriageway is shown in Appendix 8.

Summary

Health and safety law requires that employers carry out risk assessments for all the hazards associated with the work they carry out in conducting their undertaking.

The assessments must identify both hazards associated with the actual building and premises as well as any machines, materials, substances, chemicals and methods of work.

The assessments should identify the significant risks. Trivial risks can usually be ignored.

Ensure the overall assessment covers all areas and does not leave out any which could have high risks, such as maintenance, storage and catering.

Make sure you include everyone. That means those who work outside normal hours, such as shift workers, cleaners, contractors, security staff and all visitors.

Where risks cannot be eliminated, introduce safe systems of work to control the risks.

Record your findings.

Draw up procedures, company rules and train everyone in their use.

Carry out monitoring and auditing to ensure procedures are being followed and safe systems of work are in place.

Review your assessment from time to time and feed back any significant findings.

Finally, if you have carried out your assessment and produced a thick file which then sits on a manager's shelf waiting for an inspector to call, you have *failed*, and the assessment has benefited no one. Risk assessment is all about providing a means of solving practical health and safety problems at the 'sharp end', not a theoretical exercise. If assessments are properly conducted, and the resultant safe systems of work are put in place, then the level of accidents associated with engineering work will be lessened, and everyone will benefit.

Appendix 1: risk assessment (the five steps)

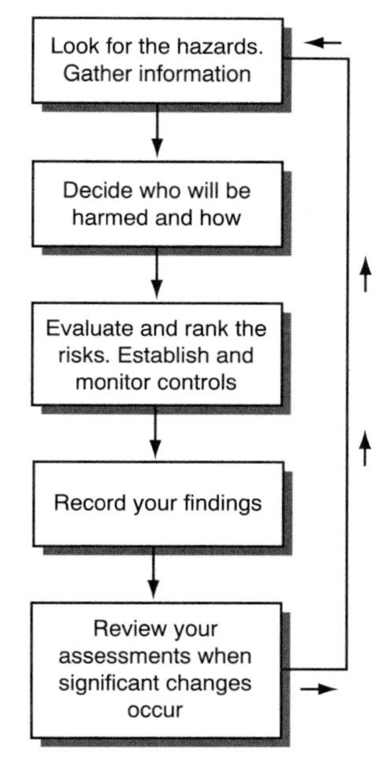

NB: Assessments that have been carried out as a result of other legislation do not have to be repeated.

Appendix 2: Cambridgeshire County Council risk assessment

The following procedure is one used by Cambridgeshire County Council to evaluate its risks and can be used as a guide.

An example of a risk assessment covering the changing of a wheel on a car (Appendix 7) is also provided as an illustration. The example shows that the concept of risk assessment is not as difficult as it often appears to people approaching risk assessment for the first time.

Purpose

To ensure that risk assessment is carried out consistently throughout the Directorate.

Scope

This procedure should be used as guidance when carrying out any risk assessment.

Principles

The principle of risk assessment is to determine systematically the hazards, the probability of harm occurring and the anticipated consequence, then to identify what controls can be used to reduce the risk or its effects.

- A *Hazard* is the potential for harm—moving traffic is a hazard, as is a trailing cable across an office walkway.
- A *Risk* is the probability or chance that harm will result—there is a risk in crossing the road or in the obstructed walkway.

We all encounter hazards in almost everything we do. Usually we keep the level of risk down by taking precautions or by applying controls—*by using a pedestrian crossing we reduce the risk in crossing a road.*

At work the law requires that employers use the principle of 'so far as is reasonably practicable' to ensure health and safety. We need to have a method of evaluating what the risks are and the best methods of reducing them. With limited resources, a system of deciding priorities in dealing with the risks must be used, and risk assessment is part of that system.

This notion is not new. It was introduced with legislation for controlling lead and asbestos at work, but was especially emphasized in more recent regulations such as the Noise at Work and Control of Substances Hazardous to Health Regulations. Previously risk assessments were carried out invisibly in our heads, and were therefore not systematic. However, regulations introduced in 1993 mean that risk assessment must now be carried out visibly, and a record kept of what has been considered and the conclusions reached. Certain activities such as manual handling, working with VDUs and using personal protective equipment are required to be assessed under their own regulations, along with tasks involving substances, exposure to noise, etc.

Method of assessment

Assessment is a tool to enable prior planning and proper allocation of resources for accident prevention.

Some highly scientific ways of carrying out risk assessments have been developed, but many of these are impractical for tasks that involve people more than machines.

The following system is basic, but its simplicity means that it is easy to use and as accurate as can be expected when assessing the human element.

A record of assessments may also be helpful as evidence of compliance with safety regulations, and of ensuring the employees' health, safety and welfare as far as possible.

Introducing controls

If significant risks are identified and you need assistance in determining which controls are possible, practical and affordable please consult the Directorate's Health and Safety Adviser (DHSA). You may also identify any training requirements.

Carrying out assessments: the practical steps

(1) Identify the different areas of activities in your unit and allocate responsibilities for carrying out the assessments to a competent group of people.
(2) Break down the tasks or jobs into similar categories and carry out an initial assessment on each.
(3) Identify the hazards for each category.

Those hazards which are insignificant or part of everyday living can be discounted.

(4) Now evaluate the risks using the matrix provided in Appendix 5, multiplying likelihood by the severity which can be expected.
(5) Consider any existing controls or precautions which are applied; this can be done after considering the 'raw' risks, as if no precautions exist, or you may prefer to include controls in step 4, and to determine residual risk, the degree of risk remaining if those precautions are taken.
(6) Using the table in Appendix 6 provided as a guide, decide if anything

further needs to be done to reduce risk. You should now be able to see the priority areas and also those where no further action is necessary.

(7) Identify which additional measures can be used to reduce risk and consider their effectiveness. Use the hierarchy of controls given in this guidance.

(8) Determine any specific areas where more in-depth assessment is required—perhaps those activities already mentioned as having individual assessment requirements. This might mean using a checklist approach for the use of chemicals, VDUs or manual handling.

(9) Ensure that you retain a record of the assessment process, and establish a good monitoring and review system. You need to communicate the results of the assessment to those affected by it, and arrange for any training that may be needed.

Basic hierarchy of safety controls and precautions

Eliminate Remove the hazard.

Isolate or contain Create a barrier between the hazard and people.

Safe system of work Use a defined system of working, following safety procedures designed to reduce risk.

Training, instruction and information Essential and should be combined with supervision. Where this is not practicable, the level of training or instruction must be even higher to compensate. Information may include warning signs or notices.

Personal protective equipment or clothing To be used as a last resort or in combination with other measures. Remember, it must be suitable for both the task and the person and its effectiveness depends on it being properly used. The degree of supervision applied will be relevant.

See Appendices 6 and 7.

Health and safety method statements

Health and safety method statements are particularly useful in establishing how a contractor is going to overcome the risks he has identified for a certain project.

If you need to know what methods he will use it is essential from a health and safety point of view that the contractor provides you with a method statement that covers exactly how and in what manner health and safety risks will be covered. When asked for a method statement contractors will often provide a standard technical version which lists the risks, but does not show in actual practice how they will be overcome.

The following example shows a method statement for the installation of some bridge beams. It combines both the technical and practical aspects of the job, and provides an illustration of what is needed for the installation of beams for a small rural bridge, off the highway, carried out by an experienced contractor, using experienced workers, both his own and the crane hire company's. Just before work starts an addendum is added: this can be discussed, and then if everyone is in agreement, and knows what their allotted task is, the risks will be reduced to a minimum, and the work can take place. A comprehensive checklist is provided as a guide in Appendix 4. As in the following example the complexity of a method statement needs to match the complexity of the task.

Example of a method statement

METHOD STATEMENT

For the .... Bridge at .... for Cambridgeshire County Council.

Additional information on 'METHOD STATEMENT' to place bridge beams

1.1 Beams are to be delivered to site on 13.1.95. by articulated lorries which will be reversed down .... The five internal beams will be on three lorries. The external beams will be on two lorries, spaced at times to suit the placing on site.
1.2 A leaflet drop to affected householders will be made on 10.1.95. explaining the need to keep the route clear (copy attached).
1.3 Cambridgeshire Constabulary has been notified and has agreed to put out 'No waiting' cones along the route from the main road to the site.
1.4 All beams to be checked before offloading.
2.1 It is intended to place beams using a 90 tonne mobile crane commencing with the five intermediate beams.
2.2 The centre of the abutment is to be set out and the spacing of beams to be marked leaving the 10 mm nominal gap between beams. The bearing strip is to be laid onto the abutment both sides.
2.3 The crane will stand on a prepared hardcored pad behind the abutment and be able to lift the beams from the lorries directly into place. It may be necessary to offload and stack two beams to allow continuity for delivery purposes. The beams are to be lifted with min 10 m long lifting chains/strops which give the minimum requirement of 45° from vertical from the lifting points cast into the beam.
2.4 The five internal beams will be placed.
2.5 The reinforcement will be fed through the holes in the beam as shown on the drawing.

2.6 The edge beams require the threaded U-bars turned into the coupler before they are placed.
2.7 Filler foam to be inserted between adjacent beams to prevent grout loss when concreting deck.

ADDENDUM TO METHOD STATEMENT

For ....

1.1 We have taken instructions from .... Crane Hire on their requirements for a crane pad. This will be constructed in the next two days and inspected for approval by a .... representative on Thursday 12 January.
1.2 There will be a minimum of five men present during lifting. Two operatives for each beam end and an experienced banksman controlling the crane. Ropes to be attached to the beam ends to steady any swing and used until 'hands on' can be achieved as the beam is finally placed.
1.3 The chain brothers* will be passed through the lifting eyes cast into the ends of the beam.
1.4 Lorry movements will be controlled to ensure that any lorries waiting will be parked in the layby on the main road and only one lorry will be allowed in the lane at any time.
1.5 On completion of the edge beams a temporary handrail to be erected from scaffolding on each side and a barrier across the bridge left after working hours.

* Two-leg chain sling.

Appendix 3: checklists

Risk assessment checklist 1—information

The six main areas that you need to gather information on to be able to apply the 'five steps':

(1) the various activities of the workplace
(2) the materials and/or substances used
(3) the plant and equipment used
(4) the details of the workplace
(5) the people who are or may be involved
(6) any existing procedures.

Risk assessment checklist 2—the process

The stages are:

- ☐ identify the hazards
- ☐ assess the risks
- ☐ eliminate or introduce controls
- ☐ identify residual risks
- ☐ provide information, instruction and training
- ☐ monitor compliance and reassessment.

Appendix 4: method statements—checklist

(1) Method statements are required for all types of work known to be hazardous and in other areas specified in the tender documents or where HSE documents indicate a need for them.

(2) This will include:
- demolition
- excavation
- work in confined spaces
- steel erection
- cladding work
- roof work
- work on high-voltage electrical installations
- hot work
- any other area where a policy statement or the construction manager indicates a need for them.

(3) Method statements should be specific to the site although they may include reference to standard or generic method statements.

(4) Method statements should be followed. Variations on the method statement should be approved by the contractor's construction manager before their implementation. Such amendments should be approved in writing.

(5) Method statements should give details of:

(i) name of the contractor's representative responsible for ensuring compliance with the method statement

(ii) risks arising from the work proposed

(iii) safe means of access to and egress from the site

(iv) arrangements for a safe place of work on site

(v) plant and equipment to be used, including size, weight or capacity limitations and, where appropriate, the test or examination certificates

(vi) training, competence and, where appropriate, the certification of plant operators

(vii) specific hazards expected

(viii) measures designed to control those hazards, controls on the hazard or the use of Personal Protective Equipment
(ix) supervision and checking to ensure that work is progressing safely
(x) routine inspection that will take place on site
(xi) precautions or limitations on the activities of others in the vicinity and the periods where these precautions apply.

Appendix 5: risk assessment tables

Severity (consequence)

Rating	Hazard effect
5	Death or permanent disability
4	Serious injury: long-term sickness
3	Three-day injury: temporary disability
2	Injury requiring medical attention or causes work restriction
1	Minor injury, e.g. bruise, abrasion

Likelihood (probability)

Rating	Probability
5	Will invariably happen: almost certain
4	Highly probable
3	Possible: feasible
2	Possible: might happen
1	Remote possibility/negligible

Source: Health and Safety Advisory Group, Cambridgeshire County Council.

Appendix 6: likelihood

		Likelihood (probability)				
		5	4	3	2	1
	5	25	20	15	10	5
	4	20	16	12	8	4
Severity (Consequence)	3	15	12	9	6	3
	2	10	8	6	4	2
	1	5	4	3	2	1

Factor (severity × likelihood)	Action
16–2	Unacceptable risk requiring immediate action
10–15	Risk reduction high priority
6–9	Medium risk: action as soon as possible
3–5	Low priority: further reduction may not be feasible or economical
1–2	Low risk: no further action required

Source: Health and Safety Advisory Group, Cambridgeshire County Council.

Appendix 7: the risk assessment form

Risk assessment form

Activity:

Hazard	Hazard effect	Severity rating	×	Likelihood	=	Risk	Minimize risk by	Residual risk

Final assessment: Overall risk:

Source: Health and Safety Advisory Group, Cambridgeshire County Council.

Risk assessment form

Activity: Changing a wheel following a puncture

Hazard	Hazard effect	Severity rating	×	Likelihood	=	Risk	Minimize risk by	Residual risk
Spare wheel difficult to remove from boot	Strained back	4		3		12	Constraint of vehicle design. Take care	4
Jack may sink in soft ground	Car might topple causing crushing injury	5		3		15	Drive car slowly to firm ground	5
Darkness and rain means it may be difficult to be seen by other drivers	You are run over and receive major injury	5		4		20	Carry warning triangle, use hazard warning lights, wear light coloured clothing	5
Wheel nuts not correctly tightened after fitting	Car unstable and crashes causing major injury	5		3		15	Check nut tightness	5
Final assessment: Wheel can be changed ensuring precautions are in hand								Overall risk: 5

Source: Health and Safety Advisory Group, Cambridgeshire County Council.

Appendix 8: generic risk assessment/health and safety plan

All current health and safety legislation and regulations apply to this document and Planning Supervisors, Designers and Contractors, Principal and Subcontractors, must understand how this applies to them and must comply with the requirements.

Activity	Risks	Potential outcome
Emergency pothole repairs	(1) Working in carriageway and being hit by fast or slow moving vehicles	Physical injury/disablement and possible fatalities
	(2) Working in difficult locations, i.e. visibility problems	Accidents caused by moving machinery, i.e. crushing injuries by roller
	(3) Moving machinery, i.e. roller or wackerplate	
	(4) Noisy machinery, i.e. compressor	Disturbance to residents
	(5) Moving in and out of both stationary and moving vehicles	Disruption to traffic
	(6) Working with hot materials	Damage to vehicles and property by hot material

Additional specific risks (list below)

....

....

....

Health and safety recommendations (safe system of work)

(1) Traffic management to comply with the County Council drawing No. and should be agreed with the Client prior to work commencing on site. Special arrangements may be required for more difficult site, i.e. dual carriageway, principal roads.

(2) Machinery to be checked and maintained regularly. Staff must be trained to be competent to use equipment. Check certificates and maintenance records.

(3) Personal protective equipment must be provided by the principal contractor and must be worn by operatives, i.e. head/ear/eye and breathing protection should be worn when appropriate. Hand and foot protection for all of hot materials. High-visibility jackets to be worn at all times on site. Check personnel on site.

(4) Hot materials should be contained and stored safely.
(5) Vehicles to be fitted with flashing warning beacons.
(6) All vehicles must carry a fully stocked first aid kit. Check.

To be returned with contract documentation

(1) Do you agree to comply with client Health and Safety recommendations? YES/NO
(2) Have you considered if there are any additional specific risks? YES/NO

Contractor's response: ....

Additional specific risks: ....

Health and safety proposals to overcome the risks: ....

The Risk Assessment and Health and Safety Plan informs the contractor of the foreseen significant risks and the minimum measures including traffic management to control the risks as far as reasonably practicable. *It does not relieve the principal contractor of a responsibility to carry out risk assessments for the work and to satisfy themselves that they have all the information needed to carry out the works safely as it may affect employees or others, and to develop and co-ordinate the Health and Safety Plan, including welfare provision.* The principal contractor shall provide details of their technical and managerial approach for dealing with the above risks and notify the client of any additional measures proposed at a pre-commencement meeting. Work must not start until the client agrees.

Name and signature of Client Assessor: .... Date: ....

Name and signature of Contractor's Assessor: .... Date: ....

Date passed to Contractor: ....

Date returned to Client: ....

Date issued: .... Issue No. 1 ....

Original source: Buckinghamshire & Warwickshire County Councils.

Bibliography

Engineering Council (1993). *Guidelines on Risk Issues*. Engineering Council, London.
Engineering Employers' Federation (1993). *Practical Risk Assessments*. EEF, London.
Health and Safety Executive (1991). *Successful Health and Safety Management*. HSE, Sudbury, HS(G)65.

Health and Safety Executive (1992). *The Management of Health and Safety at Work Regulations 1992. Approved Code of Practice.* HSE, Sudbury, L21.

Health and Safety Executive (1994). *Five Steps to Risk Assessment.* HSE, Sudbury, IND(G)163L.

Health and Safety Executive (1996). *A Guide to Risk Assessment Requirements.* HSE, Sudbury, IND(G)218(L).

Local Government Management Board (1994). *Assessing Health and Safety Risks in Local Authorities.* Local Government Management Board, London.

6. Personal safety

Elizabeth M. Bowman, *Mabilis, Shipston-on-Stour*

Introduction

With authority should come responsibility. The further up in your organization you progress, the more legal and moral responsibility you will carry for those reporting to you. This is because managers who control resources have the most effect, for good or ill, on safety performance.

At whatever stage you are in your career, you have a duty to care for yourself. This too makes sense, as you are in a better position than anyone else to observe daily activities affecting your safety or health.

This chapter is aimed at raising awareness of some of the key risks in construction work. It addresses the main duties in law on the individual, looks in some detail at construction legislation which gives minimum standards for safe and healthy working, discusses some strategies for working alone or in other high-risk situations and finally addresses emergencies.

Duties on the individual

The Health and Safety at Work etc. Act 1974 puts general duties on employers to their employees and to others affected by their acts or omissions. It also puts general duties on employees 'to take reasonable care for the health and safety of himself and of other persons who may be affected by his acts or omissions at work' and further, to co-operate with his employer in matters of health and safety. As an employee these latter duties self evidently affect you, but you should also remember that you may be acting on behalf of your employer if you are in control of your work or the work of others. While it is more common for employing organizations to be prosecuted for breaches, it is not necessarily the case that individuals are immune. Moreover it is unusual for career prospects of individuals to be enhanced by acts or omissions which have led to corporate prosecutions! The Management of Health and Safety at Work Regulations 1992 also impose overarching duties. These Regulations bear reading as a sound management document. In terms of health and safety their essence is to demand risk assessment, control and proactive management strategies. Since the assessment is an absolute requirement you will have some duty to be involved in the process, even if it is only to obey instructions relating to health and safety. Remember that if you have control over how you work, you are likely to carry at least shared responsibility with your employer.

A great deal more information is given on law and risk assessment in other chapters.

Information sources

Of course, in order to assess risk it is important to understand what risks are likely to affect your activities and determine your safety. There are several sources to which you can make reference. Each year the Health and Safety Commission publishes national accident statistics. Simple reference to the summary sheet contained therein shows that falls from height of people or objects on to people are still the major killers, while manual handling and slips, trips and falls on the level are causing extremely high numbers of major accidents. Such information gives factual statistical guidance on both the severity and probability of accidents, which may assist in risk assessment. For instance, working at heights over 2 m should always be considered high risk until adequate controls are in place. Anyone involved in activities such as roof inspections or structural surveys should be aware of this and take the appropriate action to reduce risk.

A rich source of information on risks is to be found in the legislation. Health and safety law is generally made in response to unacceptably high accident or ill health figures. Thus, since there is specific construction legislation covering specific activities, you can be reasonably confident that these are dangerous tasks. Such legislation should generally be taken as a minimum standard. You can find reference to requirements relating to your activities in libraries, from your company safety department or from the Health and Safety Executive information line. In addition to the above it will be expected that you keep up to date with health and safety management developments in your profession in the same way you keep up to date with technical and other subjects. You should be aware of guidance notes and instructions produced by your professional body. In order to provide information about specific construction risks the following section of this chapter looks at the Construction (Health Safety and Welfare) Regulations 1996. If you are familiar with the requirements of the Construction Regulations of the 1960s you will only need to take a cursory look at the 1996 regulations. You should be aware, however, that there are some additions, particularly relating to traffic routes and plant safety in general, and to fire precautions. You should also be aware that lifting operations are not included as these are being pulled together from across industry into a separate set of Regulations.

Construction (Health, Safety and Welfare) Regulations 1996: overview

Introduction

The Construction (Health, Safety and Welfare) (CHSW) Regulations 1996 express the minimum requirements for construction safety management in practice and complement the Construction (Design and Management) (CDM) Regulations 1994. A general knowledge of both sets of regulations is necessary to understand good practice in construction. In addition other regulations apply to specific high-risk tasks such as crane movements, working with asbestos, and manual handling and use of chemicals. The list is extensive. The fact that such specific legislation exists should alert you to the fact that such activities are considered high risk, requiring legislative control. A clear understanding of the CHSW Regulations will go a long way towards ensuring the safety of the individual.

The CHSW Regulations have no supporting approved code of practice. You are therefore obliged to refer to the Regulations themselves or to guidance material. A particularly highly recommended document is HS(G)150, *Health and Safety in Construction* (Health and Safety Executive, 1994). Anyone working in construction, at whatever stage, should be familiar with this document. The Regulations themselves are necessarily written in legal terminology, and care must be taken in interpretation of detail.

In terms of safety management the CHSW Regulations are on the third tier below the Health and Safety at Work etc. Act 1974 and the Management of Health and Safety at Work Regulations 1992, and they must be read in conjunction with these regulations.

Duty holders

The CHSW Regulations change the emphasis of duty holders slightly, and it should be noted that there is no longer a requirement for a contractor to appoint a site safety supervisor, but rather to manage safety. Under the new Regulations everyone has some degree of duty. The level of duty and accompanying liability is in proportion to the level of authority. It is the controller of the individual and the controller of the work who carry the main responsibility. Thus the key duty holders are the employer, the self-employed, the controller of the site and the employee (because he has control over the way he works). In particular, Regulation 4(4)(b) requires that anybody working under the control of another person reports any defect which may endanger health or safety. This duty will fall on client's representatives on site as well as on others. You are advised that while it is imperative that danger is reported, care should be exercised in reporting every single breach of legislation. This is because repeated reporting of minor infringements can have three negative effects:

(a) a smothering of major issues in a mass of minor ones
(b) deterioration of site relations
(c) confusion over responsibilities for safety.

You should be looking out for significant danger. On most construction sites the key issues are:

- falls from height of personnel
- falls from height of objects on to personnel
- falls into excavations
- collapsing of excavations
- buried services
- overhead power lines
- vehicle movements (particularly where vehicles and pedestrians are using the same area)
- stability of plant on slopes.

It is important to remember that the main responsibility for the health and safety management of construction work during the construction phase of a project still rests with the principal contractor.

Case study

After signing a motorway contract the Engineers' Representative made several reports a day to the Principal Contractor in relation to minor safety infringements. Some of the complaints were based on incorrect interpretation of the legal requirements and good practice. The contract was running behind programme. The Agent became so irritated with time wasted following up the minor infringements and incorrect complaints that when major dangers were pointed out he had become used to ignoring such notifications. The police and the Health and Safety Executive had to attend site over the weekend. Senior management for both parties became involved.

The specific exemptions to Regulation 4 are in relation to Regulation 22 (relating to welfare) and Regulation 29(2) (relating to inspections). These are dealt with under those Regulations.

The Regulations in overview

The Regulations expand upon the requirement of Section 2(2)(d) of the Health and Safety at Work etc. Act 1974 in relation to access. The requirements are more clearly expressed in that they do not simply require safe access, egress and a safe place of work, but also demand, so far as is reasonably practicable, sufficient working space and suitable arrangements of that space.

Falls from height and the use of scaffolds and ladders are dealt with in the extensive Regulation 6, which generally requires suitable and sufficient steps to be taken to prevent, so far as is reasonably practicable, any person falling. It then extends this in detail to specific cases and introduces the changed scaffolding requirements expressed in Schedule 1 and the more general requirements of Schedule 2.

It can be seen from a study of the Regulations that there is an absolute requirement to provide physical restraint where somebody can fall a distance of 2 m or more. The hierarchy here is guard rails and toe boards, personal suspension equipment and, finally, fall arrest systems.

Regulation 6 of the Construction Regulations carries on to allow for ladder work. Interestingly, 6(5) specifies that ladders shall not be used either for access to or egress from a place of work unless it is reasonable to do so. This judgement must be taken against the nature of the work carried out, its duration and the risks to the safety of any persons using the ladder. Note too that there are slight changes to the requirements for use of ladders expressed in Schedule 5. Ladders of 3 m or more should be secured if it is practicable so to do, and where it is not practicable the ladder must be footed. The previous requirement that was explicit in the extension of the ladder beyond the stepping off point has been removed and the requirement now is that a suitable handhold be provided which may either be the extension of the ladder or something similar such as adjacent scaffolding. Ladders which rise a vertical distance of 9 m or more should be provided with safe landing areas or rest platforms.

Regulation 6 goes on to talk about erection or installation of any scaffold, personal suspension equipment or fall arrest system. The key issue here is that supervision must be *competent*. Industry has judged that the Construction Industry Training Board (CITB) scheme for scaffolders should be recognized and this is generally accepted as demonstration of competence since it demands both training and experience. All scaffolders should carry a ticket and validity can be checked with the local CITB office. The requirement for toe boards does not stand for stairways or platforms which are part of a scaffold and where there is no material or substance kept or stored.

As a footnote to Regulation 6 it may be helpful to be aware that if the work is not 'construction', then these Regulations do not apply. For all cases where this is true the Workplace (Health, Safety and Welfare) Regulations 1992 are likely to apply. (Regulation 13 addresses requirements relating to falls and falling objects.) This distinction may be important commercially but remember that assessment and control of risk will still be an absolute requirement.

Steps must be taken to prevent any person from falling through any fragile material. Note that this requirement is absolute. The detailed requirements in effect consider the fragile material to be an open space, in that if anybody can fall 2 m or more, platforms and other supports must be provided. Where individuals are working near fragile material, guard rails, coverings or other similar means must be employed for preventing so far as is reasonably practicable, anyone from falling through that material. Further to this, warning notices must be affixed at the approach to any fragile material. Designers should take note of this if there are fragile roofs that will require later maintenance. It is advised that where such roofs are incorporated in any structure, information relating to them should be included in the health and safety file produced in response to the requirements of the CDM Regulations.

Falling objects are the third most common cause of death at work. The Regulations call up Schedules 1 and 2, which talk about guard rails, toe boards and working platforms. Designers may need to consider this requirement if they are specifying the order of work for construction in the pre-tender health and safety plan under the CDM Regulations. Those with responsibility on site will need to consider whether it is possible to avoid people working above others. Those responsible for site supervision will need to take particular care of material storage on site where this is at height. There is an absolute duty not to throw or tip material from height where it is liable to cause injury to any person. Those in charge of demolition and dismantling work should take particular note of this since the requirement is absolute.

Regulation 9 requires that all practicable steps be taken to prevent danger to any person from the instability of any new or existing structure including excavations. Temporary works erected to maintain such stability must only be erected or dismantled under competent supervision.

Similarly, demolition and dismantling can only be planned and carried out under the supervision of a competent person. The planning and execution must be carried out in such a manner as to prevent, so far as is practicable, danger to any person.

Explosive charges can only be used if suitable and sufficient steps have been taken to ensure that no person is exposed to risk of injury. Compare this to the requirements under the Health and Safety at Work etc. Act 1974 and the Management of Health and Safety at Work Regulations 1992.

In relation to excavations, it is of particular note that the previous requirement to provide shoring, to batter back, or otherwise make safe excavations of a depth greater than 1·2 m no longer exists. Instead, the requirement is to prevent danger in relation to accidental collapse. The requirements relating to shoring are that where it is necessary to prevent danger, the excavation shall be sufficiently supported. Installation of such support shall only be carried out under the supervision of a competent person.

Regulation 12 goes on to demand that where it is necessary to prevent danger, people, vehicles, plant, equipment, earth or other material must be prevented from falling into the excavation. Where the collapse of an excavation would endanger any person no material, vehicle, plant or equipment shall be placed, or moved near to that excavation.

On a slightly separate issue, this Regulation also demands that suitable and sufficient steps have been taken to identify, and so far as is reasonably practicable prevent, any risk of injury from any underground cable or other underground service.

Cofferdams and caissons must be suitable in design and construction and properly maintained. The construction, installation, alteration or dismantling of the same must only take place under the supervision of a competent person.

The hierarchy of control for work near water or other liquid is prevention of falling, minimization of risk of drowning or provision of suitable rescue equipment. The Regulation also requires the safe transport of any person conveyed by water to or from work and lays out minimal general standards for any vessel used for carrying people. Such a vessel shall be suitably constructed, properly maintained, controlled by a competent person, and not overcrowded or overloaded.

Regulation 15 is a new requirement concerned with traffic routes, and demands that construction sites be organized in such a way, so far as is reasonably practicable, to allow the safe movement of pedestrians and vehicles. In particular, where there is a door or gate separating pedestrians from traffic routes, it must be possible for pedestrians to see any approaching traffic. Protection can be by sufficient separation between vehicles and pedestrians or other means for protection of pedestrians, together with effective arrangements for warning any person liable to be crushed or trapped by any vehicle.

Loading bays must have at least one exit point for the exclusive use of pedestrians and if it is unsafe for pedestrians to use vehicle gates, a separate door must be provided for pedestrians in the immediate vicinity.

Traffic routes must be, so far as is reasonably practicable, free from obstruction and with sufficient clearance. Where this is not reasonably practicable, suitable and sufficient steps must be taken to warn the driver of any approaching obstruction or lack of clearance. It should be noted here that vehicles operating in one position may have adequate clearance but in a different position may not. Where this is the case particular care will need to be taken. A classic example is the movement of plant under overhead cables or through gantry structures in an elevated position.

Traffic routes must be suitably signed if this is necessary for health and safety. Note that if it is obvious that a route is for traffic, perhaps because it is a road with pavements, to tell staff it is a road is superfluous. Common sense applies.

Regulation 16 refers to doors and gates and is another new requirement. It demands that where it is necessary to prevent injury, any door, gate or hatch shall be fitted with a suitable safety device. These will include such devices as something to prevent sliding doors, gates or hatches coming off their tracks, something to prevent upward opening doors, gates or hatches falling back, something to prevent powered doors, gates or hatches causing injury by trapping, something to allow the manual operation of powered doors unless they fail to safety. It must be noted that this Regulation does not apply to any mobile plant or equipment.

Regulation 17 is yet another new Regulation concerned with vehicles. It primarily demands that suitable and sufficient steps shall be taken to prevent or control the unintended movement of any vehicle. Where movement is intentional the person controlling the vehicle must be able to provide effective warning to any person who might be at risk. Any vehicle being used for construction work must be loaded such that it can be driven, operated or towed safely. Riding on vehicles except in specified safe places provided for that purpose is *absolutely and specifically forbidden*. While vehicles are being loaded or unloaded with loose material, no one shall remain on the vehicle unless they have a safe place of work. Vehicles used for excavating or handling must be prevented from falling into any excavation pit, water, embankment or earthwork. Where tractor vehicles are used, suitable plant and equipment must be provided for replacing derailed vehicles.

There is a general requirement to prevent risk of injury from fire or explosion, flooding or asphyxiation, or 'confined space' problems. Note that many locations on a construction site are dangerous in this manner although space is not constricted. The Abbeystead explosion is a good example of this, as are many trench and manhole asphyxiation disasters.

There is a new requirement that at all times on construction sites emergency routes and exits must be provided, maintained clear of obstruction and given emergency lighting where necessary. Such routes must also be signed appropriately. It is of particular concern to those working in sites such as inner city redevelopments, process plant shutdowns, etc., where the consequences of fire can be catastrophic.

There is also a requirement to develop, communicate and practice emergency evacuation where necessary, and to provide adequate fire detection and fire-fighting equipment. Welding, hot cutting, disc cutting, etc., which may give rise to particular risk of fire must be managed by suitable additional instruction.

Suitable and sufficient welfare facilities should be provided or made available, so far as is reasonably practicable. For many small sites where the work is of short duration, the use of nearby public conveniences can be construed as appropriate. In general, however, Schedule 6 requirements should be applied so toilets, washing facilities, rest rooms, etc., will be the norm.

An adequate supply of drinking water must be provided together with a fountain facility or cups. Accommodation must be provided for storing clothing that is either not worn at work or specifically worn at work. Changing facilities must be provided where special work clothing is required. Rest facilities that are suitable and sufficient must be provided.

Fresh or purified air must be provided at every workplace and there must be effective warning devices where plant providing the same fails.

The requirements on temperature and protection from adverse weather are not specific. For instance, the Regulations require that suitable and sufficient steps shall be taken to ensure, so far as is reasonably practicable, that the temperature at any indoor place of work is reasonable. Note here that this refers to construction indoors. Other indoor activities will be covered by the Workplace (Health, Safety and Welfare) Regulations 1992. It makes sense to ask yourself whether the conditions in which people are asked to work are those with which you would be comfortable.

Protection from adverse weather can be judged following the consideration of any protective clothing or equipment provided.

Lighting must be suitable and sufficient and, by preference, natural light. Any artificial light must not be coloured so as to affect the perception of signs or signals provided for health and safety, and emergency lighting must be provided where necessary for reasons of health and safety.

Construction sites should be kept in good order and in a reasonable state of safety. The perimeter of a construction site must be identified by suitable signs. No timber or other material with projecting nails may be used or allowed to remain in place where it may a source of danger. This last requirement is absolute and means that when anybody suffers a penetrating injury from a nail on a construction site, then there will almost certainly have been a breach of the law. In any case good housekeeping on site is a reasonable indicator of good practice in general, both in terms of safety and other technical and commercial matters.

Plant and equipment must be, so far as is reasonably practicable, safe, suitable and sound. Further, it must be used and maintained so that it is, so far as is reasonably practicable, safe and without risks to health.

As is the norm for modern health and safety legislation there is a specific requirement for training. The full text of the Regulation is 'Any person who carries out any activity involving construction work where training, technical knowledge or experience is necessary to reduce the risks of injury to any person shall possess such training, knowledge or experience, or be under such degree of supervision by a person having such training, knowledge or experience, as may be appropriate having regard to the nature of the activity'.

Note that a prosecution under this particular Regulation may well be difficult to refute and if you are a manager you are advised to keep careful records of staff training. Individuals should keep their continuing professional development (CPD) record up to date.

Regulations 29 and 30 relate to statutory inspection and reporting. Any working platform from which a person could fall 2 m or more, any excavation that is supported for reasons of safety, or any cofferdam or caisson, must be inspected at specified intervals and reports made of the same. Note that reporting used to be on Forms F91 and that such forms will still be needed for lifting appliances and lifting gear. Other records and reports can now be in any format so long as they contain specified information. Such reports must be retained on site for at least three months and are one of the prime methods that client's representatives on site can use to carry out a check that working places are safe for their staff. Resident engineers, resident architects, etc., who need to access scaffolding or enter supported excavations, cofferdams or caissons should routinely ask to see the statutory report. A summary of the timing of inspections is at the back of HS(G)150 (Health and Safety Executive, 1994).

The final Regulations deal with exemption certificates, extension outside Great Britain, enforcement in respect of fire, modification of other acts and instruments, and revocation of certain instruments.

Working alone

Some of the situations in which you may experience lone working problems are surveying, condition surveys and inspections, being the sole company representative on a site, working overseas or visiting clients at their place of work. It is not the intention here to apportion liability but rather to point out some strategies for reducing danger.

First of all it is to be remembered that lone working should always be considered high risk. If it is necessary, for commercial or other reasons, to demand that staff work alone, risks must be reduced. The key factors which must be considered are training, specific risk assessment and planning for emergencies.

Training, a legal requirement under most modern health and safety legislation, is also a requirement of most institutions. It should be relatively recent, relevant and provide realistic guidance on managing workplace health and safety. Courses are available from a wide range of providers or may be available within your organization. As a pervasive part of the Institution of Civil Engineers qualification requirements, graduate civil engineers seeking chartered status have an additional impetus to maintain an understanding of practical health and safety management. Particularly strongly recommended is the CIRIA video *Not Just an Accident* which is aimed at young construction professionals.

Risk assessment, while it can be a complex analytical assessment such as is frequently used offshore or in the nuclear industry, is more usually relatively simple. You will need to *evaluate* likely risk in terms of probability and severity of outcome and *control* that risk to an acceptable level. Training is, of course, part of that control, but as a long-term requirement is usually difficult to arrange to satisfy last-minute project safety planning. Nevertheless any risk assessment should be looking at the competence of the individual as well as the inherent dangers in the task or environment and adequate training must be considered essential. Guidance on risk assessment and the absolute legal requirement to perform such a task can be found in an earlier chapter.

In the risk assessment process you will need to consider the individual, the environment, the task, the equipment, the substances and the time with particular respect to a lone worker. For instance, will you be able to carry any necessary equipment over the terrain necessary? Is the work at night?

Procedures for emergencies are absolutely vital for lone working. There should be somebody who knows your work plan and expected time of return. Although this may feel rather oppressive it can be, quite literally, a matter of life or death in the event of accident, ill health or attack. Arrangements should be written down and clearly understood by all.

The other aspect of lone working which needs careful consideration is personal health. There are a surprising number of people with chronic illnesses such as diabetes, epilepsy or asthma who may be exceptionally vulnerable if working alone. Some organizations insist on pre-employment medical screening, others do not. For those which do not, medical confidentiality and understandable reluctance by sufferers to publicize their problems mean that the onus to control the risks is usually with the individual.

It is recommended that if you have a condition which precludes you from driving a car you should consider yourself unfit. If your disease is well controlled you will have to make a judgement for yourself. There are many temporary conditions such as influenza that should also give you pause for thought as they too can be dangerous for lone workers. Anyone taking prescription medication that advises of significant possible side effects should avoid working alone.

Violence to lone workers can be a particular problem. Remember that statistically young men are more likely to be attacked than young women. The Suzy Lamplugh Trust has the best advice and some highly recommended personal alarms. The alarms work by emitting a piercing shriek at about 115 dB. This can put an assailant off balance mentally and physically, giving you vital seconds to escape. You will need to have heard the alarm beforehand or you too may be disorientated. You must *press and go*, giving a call to action to any bystanders such as 'Call the police!'

To avoid risk you are advised to plan ahead, look confident and never assume it will not happen to you. In particular do not give or accept lifts from strangers, keep to familiar territory if you can, avoid spur of the moment decisions and changes of plan, go the long way round if necessary.

If you are working alone or as the sole representative of your company in another country you will need to plan particularly carefully. Remember that safety standards and appropriate behaviour and dress may be radically different. Assumptions can be fatal.

> *Case study*
> A water authority provided its meter readers with mobile telephones to increase their personal safety. The incidence of attacks increased, because inspectors were mugged by phone thieves. Make sure controls reduce risk.

Short duration work in hazardous areas

Frequently you will find it necessary to carry out short duration work in hazardous areas: a check in an unsupported excavation, a look at the condition of flashing details on a roof. The temptation not to take the usual precautions is severe, but reference to the Health and Safety Executive summary of fatal accidents at work shows that is just this sort of work which is most likely to result in tragedy.

Always plan. Always consider what could go wrong. Always take steps to keep yourself and others safe. Never assume it will not happen to you.

> *Case study*
> A labourer and a young engineer died during a tunnelling contract, entering an area where foul air had issued from the surrounding ground. They were just going to inspect something and did not test the atmosphere.
> Recognize potential dangers, even for short tasks.

Should you be required to work on a site controlled by others where there are major hazards you should always be accompanied by someone who is familiar with the risks and the emergency response unless and until you have been trained in such matters for that site. A study of some of the reports of accidents at process plants shows how relatively small problems can escalate into major disasters if no preventative action is taken. More detail is given in a later chapter. Remember that most accidents do not take very long to happen but the consequences can be devastating. You will need to understand the risks of the area environment in which you are working unless you are accompanied by a competent individual who can manage those risks on your behalf.

Dealing with emergencies

Proper emergency planning at any level will allocate tasks in advance to those capable of dealing with unexpected or catastrophic situations. Sometimes you may be the only one available to take charge. Whatever the situation, the first rule is, *look after your own safety first*. For you to be of any use, you must be able to think and function. The second rule is to *call for expert assistance* if possible. Only after this should you try to mitigate the problems that confront you. If you are assisting someone else who is in control, listen to instructions and carry them out calmly and quickly.

Case study
A man collapsed in tunnel workings. A colleague rushed in to pull him to fresh air. While praising the rescuer, the Health and Safety Executive pointed out that had the air still been foul there could well have been a double fatality.
Look after yourself first.

Summary

In conclusion it is important to remember that you are in the strongest position to manage your own safety. While liability may obviously become important, the law is best used as a baseline for good practice. Prevention is most definitely better than blame and compensation.

Engineering is a high-risk occupation and you will need to be alert at all times throughout your career to the hazards you may face. You will also need to draw a sensible line between experience and the overfamiliarity that can breed contempt.

Never assume. If in doubt ask or check for yourself.

- ☐ With authority comes responsibility.
- ☐ Always look after yourself first.
- ☐ Work as a team.
- ☐ Know the common dangers.
- ☐ Never make major assumptions.
- ☐ Keep up to date with best practice.
- ☐ Plan.
- ☐ Don't be persuaded to take short cuts.
- ☐ Ask. Check. Test.
- ☐ Monitor your controls.

Bibliography

Health and Safety Commission (published annually). *Annual Report and Statistical Summary*. HSE, Sudbury.

Health and Safety Commission (1992). *The Management of Health and Safety at Work Regulations 1992. Approved Code of Practice.* HSE, Sudbury, L21.

Health and Safety Commission (1995). *Managing Construction for Health and Safety. Construction (Design and Management) Regulations 1994. Approved Code of Practice.* HSE, Sudbury, L54.

Health and Safety Executive (1992). *Workplace Health, Safety and Welfare. Workplace (Health, Safety and Welfare) Regulations 1992. Approved Code of Practice and Guidance.* HSE, Sudbury.

Health and Safety Executive (1994). *Health and Safety in Construction.* HSE, Sudbury, HS(G)150.

Health and Safety Executive (1996). *Construction (Health, Safety and Welfare) Regulations 1996*, HMSO, London, SI 1996 No. 1592.

The Suzy Lamplugh Trust. *A Guide to Safer Living* and *Without Fear.* Available from 14 East Sheen Avenue, London SW14 8AS.

7. Small capital and maintenance works

Stephen Fulwell, *Consultant, Nantwich*

Introduction

This chapter is written from the perspective of an engineer wishing to undertake a small engineering task within a fully operational plant while the normal activities of the operation are in progress and while minimizing the risk to people and to the plant output.

This type of undertaking is often the normal approach within both large and small organizations as the need for increased efficiency through minimal capital investment on civil work becomes the main engineering requirement. Today the high cost of civil undertakings, which significantly dilutes the return on capital expenditure, means that additional capacity is often required within the existing structures. This may involve the expansion of current capacity by minor plant modifications in order simply to increase process throughput or flexibility. This option dictates that the project will be undertaken within the normal operating environment and that the majority of the work will involve the two completely separate activities. The two activities will be required to work together while operating independently and with totally different lines of communication and objectives. In this age of objective-driven management it is essential to ensure that the risks from conflicting activities are carefully managed through effective communications and co-ordination between the process site management and those managing the new undertaking.

In addition to the trend in capital projects it has now become the norm to operate a plant with the minimum of maintenance resources and to supplement resources through either contracted undertakings or specialist services. This use of temporary workers introduces an additional requirement in order to allow for their lack of experience of the site activities and the hazards of the work they are about to undertake. The activity may also have considerable impact upon the normal activities of the site and the regular employees whose operating practices may require temporary modification.

Recent legislation, the Supply of Machinery Regulations 1992 and the Construction (Design and Management) (CDM) Regulations 1994, has placed considerably more responsibility upon both the designer and the engineer to ensure that the wide range of health and safety requirements are fully complied with at every stage. Today, and in the future, a relatively minor plant modification will require careful evaluation through risk assessment in order to clarify the range of health, safety and business risk implications. In assessing the risk it is essential to consider the undertaking on the basis of the hazards generated and the business risks likely to arise and not the financial cost involved. Often it is the relatively minor, incidental activities which possess the greatest degree of risk to both the individual and to the financial performance of the business.

In assessing the requirements of an undertaking the engineer must not only consider the health and safety implications but also the risks to the environment and to the overall business, and have identified the appropriate risk control approaches. When undertaking any assignment within an operating unit the engineer is now required to consider a much wider range of tasks requiring significantly more hazard and procedural investigation at the planning stage and with increased control during the critical installation and commissioning stages.

The role of the manager as the client

The plant manager/engineer today clearly has a responsibility to ensure that the safety duties required of the organization are fully complied with and that the level of competence/resources appropriate to the task to be undertaken is available. In the past the subject of competence was normally considered to be resolved if an experienced contractor was appointed and defined as a specialist. Today the duties of the client in assessing the competence of the various parties involved and the need to ensure effective control and supervision throughout the activity has resulted in the need for fully documented assessments.

The most critical and often the most neglected stage of many undertakings is the preliminary concept and design stage, which if neglected often results in overspends, engineering compromises and expensive delays because adequate and effective preparation was not undertaken at the early stages. Any project, whether maintenance or capital works, in an operational environment must be carefully assessed and the implications fully understood by the client before it is cosseted or submitted to tender.

In undertaking this preliminary assessment the client/plant management must consider both the normal activities being undertaken and also the exceptional requirements dictated by the work to be undertaken and the potential implications, the 'what if' scenarios which could be encountered. These may be operational, logistic, supply chain, health and safety, environmental or business risk elements, all of which could dictate a specific course of action and influence the scope or detail of the undertaking and the manner in which the work proceeds. It is essential that this preliminary evaluation is undertaken by the client's management team prior to a detailed specification being prepared. Figure 1 outlines a simple risk assessment form which may be completed as a preliminary evaluation approach to any site activity.

Having defined and secured an outline proposal from which the scope of work, organizational needs and individual competencies required have been assessed, the client is now able to identify and resource the team appropriate to the undertaking. Competence in this situation is defined as having the appropriate knowledge, skills, attitude and awareness for the duties to be undertaken. It is also essential to identify at this stage the levels of competence or skills required by the contractors, as these may often be required to support any shortfall in the management availability and could therefore have a high priority.

Any major undertaking in a fully operational plant which involves integrated working by both employees and contractors must, at the earliest possible stage, involve a discussion among all of the parties involved with the activity and working within the associated activities. The specification identified must detail not only the machinery and civil requirements for any suppliers but also the implications for the client's operations which are to be accommodated. Special requirements in respect of a wide variety of arrangements may be required, along with exceptional insurance or operating arrangements for the duration of the work.

A strategy must be prepared by the client which enables the commercial risks to the undertaking to be managed and therefore any possible impact during the programme to be minimized. All parties must be fully aware of the programme of activity and their individual role in achieving a successful outcome. The majority of activities fail because either one or more of these key initial elements is not fully undertaken or was not adequately communicated to all of the relevant parties.

Health and safety regulations now require that construction activities are undertaken within the rules established under the Mobile and Temporary Workers Directive, 92/57/EEC, which has established a clear structure and notification procedure for the client to follow in order to comply with the duties. The wide definitions applied to the regulations in respect of the terms construction and structure in the UK often results in a wide range of site activities falling within the requirements of the Construction (Design and Management) Regulations 1994.

Activity	Hazard	Likelihood	Safeguard

Figure 1. Layout of risk assessment form

'Structure' is defined as any fixed plant in respect of work which is installation, commissioning, decommissioning or dismantling, and where any such work involves a risk of a person falling more than 2 m.

'Construction work' is the installation, commissioning, maintenance, repair or removal of mechanical, electrical, gas, compressed air, hydraulic, telecommunications, computer, or similar services which are normally fixed within or to a structure.

These regulations and the associated Construction (Health, Safety and Welfare) Regulations 1996 clearly identify the standards and procedures to be considered where a number of contractors are to be employed in a plant activity. The Health and Safety Executive (HSE) guidance document, *A Guide to Managing Health and Safety in Construction*, identifies the following stages in the development and execution of an undertaking;

(a) concept and feasibility
(b) design and planning
(c) tender/selection stage
(d) construction phase
(e) commissioning and handover

The client must first establish whether or not the requirements of the undertaking are within the extensive scope of the CDM Regulations. If applicable, a Planning Supervisor must be appointed in order to ensure the production of a pre-tender document detailing the working arrangements and the preliminary scope of the site hazards (Appendix 1). The Planning Supervisor must also ensure the production of a post project Health and Safety File (Appendix 2).

In the case of machinery installations under the Supply of Machinery Regulations 1992 the purchasing policy must be agreed with the purchasing team. This will identify the duty holder for the Declaration of Conformity and from this the location and availability of the Technical File documentation.

The client, having identified the various elements of the project, must undertake the following in respect of the ownership of the undertaking;

(a) establish the risks involved and determine the responses that the client will manage through systems and procedures and those to be passed to other organizations such as the contractors, agents, designers, suppliers, insurance or other specialists through the contracts
(b) allocate the client's responsibilities to nominated individuals/contractors and ensure that competence is established for the duties involved
(c) ensure that the full scope of risks and hazards are understood and allocated.

At this stage particular care must be taken in order to recognize that while a task may be passed to a contractor, the client retains the ultimate responsibility unless an agent is appointed with clearly specified responsibilities and the associated control. Having identified the duties to be contracted out to external organizations, it is now possible for the purchasing negotiations to begin with the various parties.

The selection of contractors and the criteria to be applied must reflect the duties to be undertaken in respect of specialist knowledge, reputation, experience and their past record in undertaking the type of work proposed. Negotiations must also verify the health and safety competence and management arrangements. Examples of the types of questions the potential contractor should be asked are contained in Appendix 3.

The client must arrange for the production of a pre-tender document and the associated rules for contractors which are to be applied for the duration of the activity and should include details of emergency and reporting arrangements. The facilities to be available to the contractor and details of any restrictions which are to be in force because of the normal undertakings must be specified. Any special arrangements must be considered at this early stage in order to prevent both parties entering into an undertaking with conflicting ways of working. For example, if the contractor is planning weekend working and the client is not, can access be maintained without the need for an interruption in the operations for a special activity?

In addition, it is essential to review jointly with the possible contractors the operational constraints and the potential consequences of a plant failure in order to ensure a full understanding by both parties of the activity requirements. The worst possible scenario is for the contractor to require an electrical supply interruption for a connection when the production process cannot be stopped. The production and complete understanding by all parties of a detailed timing plan for the duration of the activity is fundamental at this preliminary stage.

The final phase which the client must consider concerns the supervisory and inspection arrangements to be applicable for the duration of the activity. This must ensure that the allocation of control and the regular inspections by competent persons is undertaken in order to ensure full compliance with both legal and the company/site requirements. Regular inspections are an important requirement in areas where activities are being undertaken separate from the main function of the plant. Often the different groups are not able fully to appreciate the wider implications of the ways in which they work or the impact upon the other parties. Particular care must be taken where small-scale engineering activities are to be left overnight while a production shift continues as new hazards are often generated by seemingly innocent interventions by persons unaware of the consequences.

The inspection (Appendix 4), undertaken by a competent person with authority, must ensure that the standards of housekeeping and safeguarding arrangements are being maintained and that any special procedures or systems established in respect of the activity are operating correctly. These inspections must be recorded and communicated to both the contractors and the client's duty holders where external agencies are involved. Any deficiencies must be acted upon immediately and any persistent failures in arrangements reported to senior management for further action. In addition a post-start review may be undertaken (Appendix 5) by the Planning Supervisor if requested by the client in order to verify compliance with the plan.

Mechanical and electrical work within an operational environment

The installation of both mechanical and electrical components represents a significant undertaking at any premises. Work to be undertaken by internal or external resources requires careful assessment prior to the work commencing. The success of an undertaking will be determined by the planning undertaken and the extent to which potential problems likely to be incurred are anticipated and resolved

In an operational plant the most basic mechanical activity must be carefully planned and the degree of risk considered in order to ensure that the two objectives of safe production and engineering activities are successfully undertaken. At the plant level each stage of the programme must be considered beginning with the initial access and storage arrangements required if basic mistakes are to be avoided. For example, can large equipment gain access from the highway at the required times? Is a loading and storage area available if the equipment is to be held? Who is responsible for the unloading—the supplier, the haulier, the installer or the client—and has appropriate insurance cover been arranged and verified? In the event of production having to be disrupted, have the consequences of an increase in the timing been anticipated and are contingency plans available?

Within the plant it is essential to ensure that 'suitable and sufficient' risk assessments have been undertaken and safeguarding arrangements organized in order to ensure an efficient and safe programme of work. This may involve temporarily supplementing the site's normal resources in order to undertake specific duties or provide additional support. If the work is to be undertaken over an extended shift period, are the communications and co-ordinating arrangements in place in order to ensure a complete understanding? It may be appropriate at the start of the activity to bring together the parties involved, including the existing production staff, in order to ensure a common understanding.

The electrical stage of the activity in a fully operational plant is often the most dangerous element as the hazards are not normally visible. Frequently electrical drawings and circuitry have been modified over time without the production of adequate records, particularly in cases where modern software renders change a simple process. The plant electrical engineers must undertake a detailed review of the proposal, and identify the extent of the work and the safety arrangements. Often it will be necessary to identify specific cables, and confirm numbering and connection points within the control panels in order to ensure that the correct isolation has taken place. Care must be taken to ensure that no additional supplies have been missed or that accidental reconnections are not possible. The Electricity at Work Regulations 1989 Regulation 4 requires that any work activity on a system 'shall be carried out in such a manner as not to give rise, so far as is reasonably practicable, to danger'. This represents a very specific requirement and dictates particular attention in respect of safe systems of work and an installation in accordance with both the regulations and the Institute of Electrical Engineers (IEE) electrical standards.

The use of programmable electronic systems (PES) in order to control, protect and monitor systems has resulted in the establishment of two separate groups of electrician—the systems engineer and the field wiring electrician—engaged in the installation and connection of cables remote from the control centre. In many situations these two groups may be different employees or even different contractors as often the control panel may be built by the systems engineering company. In this situation it is essential to design and implement safe systems of work which ensure that comprehensive procedures exist to eliminate inadvertent live connections or incorrect numbering that could result in electrocution. The sequence of events comprising the isolation, disconnection, cable laying, connection within the panel and the final stage of reconnection through to making live must be clearly understood by all parties at every stage. Regrettably this is often the activity which follows machinery installation and is often the most hastily conceived and undertaken activity.

Following on from this is the testing and commissioning stage where a careful review of the procedure to be adopted is essential as outcomes are often not certain and the persons involved have yet to gain the experience which gives rise to competence in the activity. During this stage frequent modifications are made to the electrical control systems, timers modified, overloads adjusted, sequences modified and often reconfigured in order to eliminate the problems which arise. These modifications must be carefully reviewed and documented, any changes to the safety systems must be fully considered and the possible consequences understood. Production demands must not result in a reduction in the level of safety introduced.

The installation of associated services comprising steam, compressed air, lighting and gas supplies must also be considered in advance and the risks and regulatory requirements clearly understood. The location of these services is often of particular significance: to be hygienic they must be located off the floor, to be maintained they must be accessible. Positioning and ease of access are therefore key considerations: air lines often need draining, steam lines need 'nipping-up' and the condensate traps inspecting, electrical junction boxes will require inspection and possibly additional connections during their lifetime.

The Pressure Regulations 1989 require that welding on steam lines is undertaken by a suitably qualified person and tested/inspected by a competent person from the insurance organization who ensures that the duties required by the client are fulfilled. Any gas pipework or installation is subject to the Gas Safety (Management) Regulations 1996 and the Pipelines Safety Regulations 1996 in respect of large-scale usage on boilers and large ovens. These factors must all be considered during the planning stages of more complex installations.

The requirement to review under the Supply of Machinery (Safety) Regulations 1992 following any new installation or significant modification to an existing system may require specialist assessment by an independent organization for highly dangerous equipment. In the case of most equipment this involves a Declaration of Conformity (CE) assessment that the installation complies with the European Directive 89/392/EEC and has the appropriate CE documentation. Normally individual items of equipment will have been purchased with this approval but if the equipment is part of a 'linked assembly'—being designed and installed by the company from different sources or imported from outside the European Union—it will be the client's responsibility to assemble the Technical File (Appendix 6) and ensure CE conformity.

The final requirement is to ensure that all of the site documentation has been updated, the new risk assessments undertaken, relevant training and information provided and the management controls established. This must ensure that the additional hazards introduced have either been removed or the appropriate safeguarding arrangements established.

Practical examples of specific hazardous activities

In order to illustrate the requirement for special considerations it is appropriate to review a number of specific examples which are likely to be encountered in the workplace and constitute a small capital or maintenance activity.

Asbestos removal

This is an increasingly persistent activity with very specific legislation providing for the identification of the type, the competence of the contractor, the method of removal of the material and the ultimate disposal from the site. Normally, having identified the type of asbestos, an approved contractor is selected who is able to ensure the correct safeguarding procedures. This will contain the potential spread of the material by totally sealing off the area with airlocks that prevent the contamination of adjacent areas. In addition the persons removing the material may be required to undertake a complete change of clothes in a controlled area before they are able to leave the area. The material removed must be carefully packaged in a correctly labelled bag, stowed in an enclosed skip and the written disposal documentation authorized to a registered outlet.

In this situation the hazards generated are extensive, with all parties required to assess carefully all aspects of the activity and with effective control and safety monitoring arrangements in place. Special arrangements may be required in respect of the fire and emergency procedures if the spread of the material is not to cause a major incident in the event of a routine fire drill.

Scaffold erection

The use of scaffolding for access to high levels is a more frequent undertaking today and must be carefully considered in respect of a number of implications;

(a) selection of the scaffold contractors in respect of competence and suitability
(b) access to the area by the scaffold erector and safeguarding for existing personnel during the installation of the scaffold
(c) positioning of the scaffold may introduce new hazards or restrict vehicle and pedestrian movements
(d) provision for the regular inspection of the assembly by a competent person
(e) control of the high-level activity and any movement beneath the scaffold.

The positioning and assembly of any scaffold must be carefully considered in advance in order to minimize the risk to the employees. Normally installation requires access to the area of a vehicle with a large quantity of materials which must be assembled by a competent crew who are fully aware of the requirements and the activities being undertaken around them. Equally, persons who normally work in the area must be advised of the scaffolding purpose and the area suitably cordoned off. Under no circumstances should access beneath the scaffold be permitted during its assembly.

Use of a mobile crane within a facility

The use of a mobile crane, particularly within the premises in order to install or remove equipment, requires careful planning and selection if the risks are to be minimized. Under no circumstances should the work be undertaken without careful evaluation. Particular care must be taken when a lift involves the use of two cranes working together—as may be the case with very large assemblies—as the risks involved are substantially greater. The lift must be undertaken by a competent organization with the correct insurance cover and controlled by fully qualified staff if the risks are to be minimized.

The HSE specifies 27 questions which the organization undertaking the lift must clarify in order to undertake the activity (Health and Safety Executive, 1980), in addition the client must ensure that the appropriate safeguarding arrangements are in place prior to the lift taking place. These precautions will include the inspection of the lift company's insurance and competence documentation as well as the possible requirement to seal off and supervise the area which may be affected for the duration of the lift. Additional precautions may be required if the lift involves the travel of the crane jib over the roof of occupied premises or in close proximity to overhead power cables.

The task dictates a careful evaluation by a competent organization and an extensive dialogue between the parties involved with each party being fully aware of the changes in the duty holder and the responsibilities throughout. The installation of a steelwork assembly may be assisted if the designer is fully involved at the planning stages of the lift as they are often able to influence the degree of difficulty by adjusting either the steelwork design or the assembly criteria.

Installation of a mezzanine floor assembly

The installation of a mezzanine floor is often undertaken in order to provide additional floor space without the additional cost of building work. In undertaking this approach two key factors must be considered. Firstly, is the designer fully aware of the scope of the requirements in respect of the span and loadings involved? Secondly, can the components be safely accessed into the area and assembled while other activities are being undertaken adjacent to it?

Three organizations must work closely together in order to achieve the objective:

- ☐ the client, who must clearly specify the duty and the constraints to be applied during the installation stage—these may include size, welding and cutting and access requirements
- ☐ the designer, who has the task of designing the structure and ensuring that both the client's specification and the safe assembly requirements are fully met
- ☐ the installation team company must be carefully selected, their method statement reviewed and the detailed requirements clearly understood prior to the work commencing.

The installation stage must be carefully managed in order to remove the risk to other persons working in the area and to control the method of installation being employed by the erectors. The hazards introduced by the movement and installation of large sections of steel combined with the certainty that at least one area will not correctly fit into place dictates that effective controls exist. Whenever possible steelwork modifications should be undertaken outside the factory and away from the normal activities in order to minimize the risk of an incident associated with the cutting, welding, burning or working at a height normally undertaken. If this is not possible, the appropriate permits to work, fire precautions segregation and supervision must be initiated in order to prevent persons being exposed to additional hazards in the workplace.

Groundwork activities

The requirements for below-ground excavations normally result from either additional foundations being required or access to existing services due to additional connections or urgent maintenance needs. Current procedures dictate that before work is undertaken the location of everything below the ground must be determined/confirmed in order to minimize the risk of discovering unexpected electrical or water supply lines. Historically the electrical mains supply, the water main and the site gas line are rarely at the point shown on the site plan and are even less likely to be at the depth specified. Lengthy surveying of the affected area using modern technology is dictated, followed by the careful excavation and exposure of the relevant cables or pipework is required in order to confirm the positions.

Having identified and recorded the correct positions of the underground services, it is now important to undertake a risk assessment of the hazards and implications of the excavations for the existing operations. For example, can the hole be dug safely without risk to either the contractors or to employees undertaking their normal activities? A review of all vehicle movements through the area is essential, along with the implications of closing off the area and the method and route for the disposal of any debris which must be removed from the area. Are barriers erected which will restrict normal movement within the area?

Roofwork activities

The HSE regard roofwork as one of the most hazardous activities undertaken at all locations and it is therefore important to ensure that the correct procedures are in operation prior to any work being undertaken. The minimum requirement is a fully documented permit-to-work procedure which identifies the work to be undertaken, clarifies the method being adopted and establishes the appropriate safeguarding arrangements.

In addition the client must again be confident that the persons undertaking the work are competent and fully understand the level of risk involved, and work must not be undertaken without a thorough assessment of the scope involved and the potential risks generated both to the persons on the roof and also to those who may stray beneath them. The access route from the ground must be identified and, if required, protected, and the correct design and number of crawling boards must be available to facilitate access to the points required. Emergency procedures must be reviewed, potential high-level hazards removed or controlled and when possible a safety harness utilized. Adverse weather conditions, such as rain, snow or high winds, must be factors considered when roofwork is required, and any deterioration in the weather must prompt management action.

Communications and liaison

All of the activities outlined in this chapter place particular emphasis upon two points. Firstly the owners of the premises retain full management responsibility throughout any undertaking. Secondly effective communications and liaison across all activities involved at the location are important in order to ensure that even the most unexpected event is fully understood in terms of the potential consequences and that activity is not undertaken in isolation. Unconnected events frequently have a tendency to impinge upon each other through seemingly unconnected links and often effective communications enable parties not directly affected to anticipate outcomes.

Permit-to-work procedures and operations

In order to minimize the risk of accidents, formal written procedures are designed and implemented for specified hazardous activities. Examples of activities requiring a permit to work include the following:

- general maintenance permit
- hot work permit
- electrical isolation permit
- roofwork permit
- contractor's authorization
- entry into a confined space
- work on a pressure system

In all cases the permit is issued by a competent person who has been suitably trained and is aware of the hazards of the activity and the controls required in order to control the risks generated. They must also be aware of the implications of any associated activities and the potential consequences of a safe system of work failure. The permit should be issued solely for the day and be prominently displayed for reference by any other parties undertaking work in the area. At the end of the day the permit must be cancelled by the originator or his nominated deputy.

Permits to work should comprise the following information and be correctly documented;

(a) the permit title
(b) the permit number
(c) the job location
(d) the plant identification
(e) description of the work to be undertaken and the limitations
(f) hazard identification—those existing and those generated by the activity
(g) precautions required in order to control risk
(h) protective equipment required
(i) authorization confirming that all isolations and precautions are in place
(j) acceptance confirming a full understanding of the hazards/precautions
(k) extension/handover procedures
(l) handback of permit procedure—certifying all work completed
(m) cancellation details certifying that the work has been inspected/tested.

Permit-to-work procedures are most often associated with non-production activities; however, in a joint production/engineering activity it is essential that their use is co-ordinated normally by restricting the point of issue to a single nominated person, or for shift work to a duty holder.

Many instances are apparent where the failure to co-ordinate activities has resulted in a serious incident mainly due to the 'failure in the performance of a person associated with the system or activity' (Terry, 1991).

Maintenance shut-down activities

Many facilities schedule a routine maintenance shut-down in order to undertake the essential engineering work required to maintain optimum process efficiency. This work is often undertaken by external resources who either supplement the regular maintenance resources or possess the specialist skills required.

The programme of work must be carefully planned and controlled by a single duty holder; the practice of a number of different persons arranging various activities will result in uncontrolled activities and a high level of risk. All of the proposed activities should be advised to the duty holder who will identify the overall programme of work, the resources required to complete the task and, where appropriate, clarify the method of work being applied. From this review the duty holder will establish the extent of the risks to be encountered and the requirements for safeguarding arrangements. This may range from specific permits to work through to a routine inspection or the level of supervision required in order to maintain control of the hazards being generated during the activity. At this stage it must also be confirmed that sufficient time and resources are available in order to complete the intended programme of work.

In the case of frequent maintenance activities by external organizations it may be appropriate to undertake a specific safety audit of the contracting company and a site induction programme of their engineers if frequent visitors. However, in this situation it is essential to apply the site-recognized isolation and permit-to-work procedures and not introduce a different arrangement which confuses other parties. Upon completion of routine maintenance activities the co-ordinator must be advised of any new hazards which may have been generated followed by the inspection and testing of the work undertaken and the area involved. Any changes in the operating requirements or adjustments made which are likely to affect the operating criteria must be documented and the operating staff advised prior to restarting the equipment involved.

Conclusion

The engineer must at all times recognize the imprecise nature of health and safety: regrettably there is no single safe solution, each activity will vary across every site and must therefore be continually reassessed. 'Safety is not in itself a precise state or situation and it cannot be described in absolute terms' (Terry, 19??).

Endeavouring to achieve progressively higher safety standards through the application of the principles of risk assessment based upon hazard identification and effective management controls is the objective. In addition the application of detailed planning, effective communications and the provision of adequate time and resources will result in a successful undertaking.

The engineer has a major part to play in the establishment of safe working practices and the development of a positive safety culture within the work environment. The manner in which the task is assessed, operating standards applied and safe working systems enforced will make a significant contribution to the overall success of the business undertaking.

Appendix 1: the pre-tender stage Health and Safety Plan contents

(1) A description of the work involved.
(2) Relevant timescales including any key dates.
(3) Relevant risk assessment findings.
(4) Existing site conditions and any contamination or other hazards.
(5) Drawings of the existing site and relevant arrangements.
(6) Service details including any site survey findings.
(7) Security arrangements.
(8) Access arrangements including any time restrictions.
(9) Available working and storage arrangements.
(10) Traffic routes and restrictions to be used by the contractor.
(11) Welfare arrangements available.
(12) Impact upon the project of any site activities.
(13) Site rules for contractors to be enforced.
(14) Current and previous land uses.
(15) Structural condition.
(16) Known hazardous materials or processes.
(17) Third-party involvement.
(18) Emergency arrangements.
(19) Reporting structure and responsible persons.
(20) Communication and liaison arrangements.

This document must be drawn up by the Client who may consult with both the designers and the Planning Supervisor during the compilation of the document. The contents are to be specific to the undertaking and must demonstrate that the correct degree of planning, organization and management control has been applied. If appropriate, other specific specialist skills should be consulted during the planning process, i.e. steel designers for the steel erection procedure, lifting specialists for crane work or radiography specialists if vessels require inspection.

Appendix 2: CDM Health and Safety File checklist

Site: .... Date: ....
Project: ....
....

This checklist is produced in order to specify the contents required under the CDM, Construction (Design and Management) Regulations 1994 and may be subject to amendment under subsequent or additional regulations.

Client: .... telephone ....
Principal Contractor: .... telephone ....
Designer: .... telephone ....
Planning Supervisor: .... telephone ....

Purpose. The requirement for the Health and Safety File is in order to have available the information relevant if any further work was undertaken on the structure and must relate to the as-built final outcome and not the expectation. Any particular hazards generated by and during the construction must be clearly specified as well as any aspects identified during the progress of the installation.

Record of as-built drawings: ....
Details and location of all relevant services including depths: ....
General details of the construction methods and materials: ....
Availability of any structural calculations: ....
Specific maintenance requirements: ....
Any specialist manuals obtained: ....
Details and location of specialist utilities: ....
List of any specific hazards advised and appropriate safeguarding: ....
Details, including positioning plans and load calculations undertaken: ......
Identification of any environmental elements: ....
Name and date of inspection and approval: ....

Appendix 3: contractor selection questions

This is not a complete list but a prompt list to which specific questions should be made.

Experience

What experience have you in this type of work?
How familiar are you with the hazards involved with this activity?
What, in your experience are the particular problems you are likely to encounter?
Can you provide existing method statements or risk assessments?
Can you supply references?

Health and safety policies and practice

Do you have a written safety policy document?
What are your health and safety procedures?
Do plan to use any subcontractors?
Will you provide a safety method statement?
What safety checks do you make on equipment and materials?
Who is responsible for health and safety at director level?
Are activities inspected by a competent person or independent organization?

Training and competence

Are you a member of a trade/professional body?
How do you ensure that all subcontractors are competent?
What health and safety training do you provide?
How do you prepare subcontractors for the work involved?
What health and safety training do you provide for the site management?
How is information about health and safety distributed?
Can you provide training records for the persons on site?

Supervision

How do you intend to supervise the job?
What is the organizational structure?
How are changes in requirements and procedure undertaken?
What is the reporting procedure if a problem arises?
Are you prepared to abide by the company rules?

Enforcement action

Has any prosecution of your company been undertaken over the last five years?
Are you currently in receipt of any improvement or prohibition orders?
Have you been contacted by the Environmental Agencies for any complaints?
What is your accident reporting procedure?
What is your accident ratio in respect of the current national statistics?
How do you obtain up-to-date health and safety information?

Other questions should be added specific to the competencies required.

Appendix 4: daily inspection

Date: .... Project: ....
Site inspection by: .... Reported to: ....
Accidents reported: .... Contractor numbers: ....
All contractors registered on site: ....
Reporting structure clearly known: ....
Permits to work issued by: ....
Method statements received, reviewed and followed: ....
Chemical hazards identified and storage in compliance: ....
Accidents and near misses being reported and followed up: ....
Safe working at high level: ....

Housekeeping in the work area: ....
Warning/hazard notices in position: ....
Barriers/segregation arrangements: ....
Electrical arrangements safe: ....
Portable appliance testing and inspection undertaken: ....
Communications and co-ordination in progress: ....
Supervisory arrangements satisfactory: ....
Welfare facilities inspected: ....
Safe working practices in operation: ....
Fire arrangements satisfactory: ....
Welding arrangements in compliance: ....
Lifting equipment and arrangements correct: ....
Personal protective equipment in use and available: ....
Factory personnel advised of work programme and hazards: ....
Work in electrical cabinets safeguarded: ....

Appendix 5: CDM inspection checklist

Date: .... Project:..
Site Representative: .... Planning Supervisor: ....
Principal Contractor: .... Designer:...

Preliminary work

Pre-tender document produced and issued: ....
Contractor's competence assessed: ....
Insurance documentation checked: ....
Rules for contractors issued: ....
Design safety factors reviewed: ....
CDM notification F.10 undertaken: ....
Health and Safety Plan produced and reviewed: ....
Site arrangements satisfactory: ....
Reporting arrangements established/documented: ....
Induction arrangements and hazards advised: ....
Implications of confined space activities: ....

Work in progress

Method statements submitted/complied with: ....
Fire protection/hot-work arrangements: ....
Use of PPE correctly operating: ....
Permits to work issued and cancelled: ....
Work at height correctly undertaken (>2 m): ....

Roofwork procedures observed: ....
Welfare facilities adequate and maintained: ....
Storage arrangements reviewed: ....
Segregation arrangements for traffic: ....
Portable appliances tested/inspected: ....
Welding arrangements and certification: ....
Electrical arrangements comply: ....
Contractor's site supervision identified: ....
Communications and co-ordination effective: ....
Chemical arrangements, storage/notification: ....
Vehicle deliveries offloading and supervision: ....
Company co-ordinator identified: ....
Project engineer arrangements: ....
Lifting and inspection arrangements: ....
Contractors individually registered on site: ....
Safe working practices observed: ....
Procedure for confined space working implemented: ....
Condition of the contractor's work area: ....

Comments

....
....
....

Appendix 6: The Supply of Machinery (Safety) Regulations 1992

Date of application

All machinery first supplied or placed into service from the 1st January 1993 or subject to extensive modification or overhaul by the manufacturer.

Declaration of conformity

The procedure whereby a responsible person declares in respect of each item of relevant machinery that it complies with all of the essential health and safety requirements applying to it. The 'CE' mark is affixed to the machine.

Declaration of incorporation

This declaration, with a 'CE' mark being applied, is applicable in the case of relevant machinery which is intended for:

(a) incorporation into other machinery
(b) assembly into other machinery
(c) is not interchangeable equipment
(d) cannot function independently

states the machinery must not be put into service until the requirements of the declaration of conformity have been fully complied with and the document issued.

Certificate of adequacy

A document drawn up by an approved body to which a manufacturer has submitted a technical file and it has been verified as having correctly applied the appropriate transposed harmonized standards and contains the necessary information.

Series manufacture

The manufacturer of more than one item of relevant machinery of the same type in accordance with a common design.

Safety components

A component provided, that is not interchangeable equipment, which the manufacturer or his authorized representative established in the community places on the market in order to fulfil a safety function when in use and the failure of which endangers the safety or health of exposed persons. This is applicable from the 1st January 1997 amendment 94/44/EEC.

Technical file

A responsible person must compile or be able to assemble a technical file comprising the following information:

(a) overall drawings of the machinery together with drawings of the control circuits.
(b) full detailed drawings and calculations, test reports as may be required in order to check conformity with the essential health and safety requirements.
(c) a list of the essential health and safety requirements.
the transposed harmonized standards.
standards adopted.
any other technical specifications applied.
(d) a description of the methods adopted to eliminate hazards presented by the machinery.

(e) any technical report or certificate from a competent body.
(f) any technical report applicable to the declaration of conformity.
(g) a copy of the maintenance and operating instructions drawn up in accordance with the provision of information and in the language of the user of the machinery.

The instructions must also identify any residual hazards required to be considered.

Bibliography

British Standards Institution (1992). *Requirements for Electrical Installations, IEE Wiring Regulations*, 16th edn. BSI, Milton Keynes, BS 7671.

Chapman, C. and Ward, S. (1997). *Project Risk Management: Processes, Techniques and Insights*. Wiley, Chichester.

Health and Safety Commission (1995). *Managing Construction for Health and Safety. Construction (Design and Management) Regulations 1994. Approved Code of Practice*. HSE, Sudbury, L54.

Health and Safety Commission (1995). *A Guide to Managing Health and Safety in Construction*. HSE, Sudbury.

Health and Safety Commission (1997). *Safe Working in Confined Spaces. Approved Code of Practice, Regulations and Guidance*. HSE, Sudbury, L101.

Health and Safety Executive (1980). *Management's Responsibilities in the Safe Operation of Mobile Cranes*. HSE, Sudbury.

Health and Safety Executive (1986). *Guidance on Permit-to-work Systems in the Petroleum Industry*. HSE, Sudbury.

Health and Safety Executive (1989). *Electricity at Work Regulations, 1989. Guidance on the Regulations*. HSE, Sudbury.

Health and Safety Executive (1997). *Managing Contractors: A Guide for Employers*. HSE, Sudbury.

Horsley, D. (1998). *Process Plant Commissioning*, 2nd edn. Institution of Chemical Engineers, London.

Terry, G.J. (1991). *Engineering System Safety*. Mechanical Engineering Publications, London.

Townsend, A. (1992). *Maintenance of Process Plant*, 2nd edn. Institution of Chemical Engineers, London.

8. Managing hazardous substances in construction

Nicholas E.G. Martens, *Bechtel Ltd, London*

Introduction

What is occupational health?

In essence, occupational health is about the two-way relationship between work and health and their mutual effects (Fig. 1).

Although we tend to view occupational health as being the effects of hazards in the working environment on the health of the worker, it is just as important to recognize that the individual's state of health influences their ability to carry out the work safely.

Competence

As an engineer, your involvement in managing the hazards associated with the construction process could be key, and as such may require you to get involved, to a greater or lesser extent, depending on your role in the project. As a designer you will be expected to identify health hazards associated with your design and specification with a view to avoiding them if possible, and at the very least reducing the risks posed by them to an acceptable level. Your involvement on site in a management role may require you to get involved in the assessment, prevention and control of risks associated with hazardous substances and construction processes.

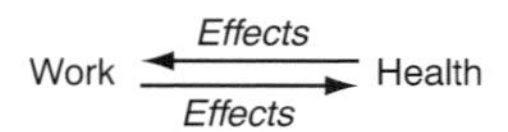

Fig. 1. A two-way relationship

Whatever your role, you will need to be competent in health and safety in order to manage hazardous substances in both the design and construction processes. That said, part of being a competent professional is the ability to know one's limitations and to get other professionals involved at the appropriate time. These may include, for example:

- occupational health doctors
- occupational health nurses

- ☐ occupational hygienists
- ☐ health and safety professionals
- ☐ toxicologists
- ☐ microbiologists
- ☐ ergonomists.

However, for the majority of engineers, being competent in relation to occupational health will mean a good understanding of the following:

- ☐ hazard identification
- ☐ risk assessment
- ☐ construction operations and processes
- ☐ prevention and control measures
- ☐ monitoring techniques.

In addition to the above, it is essential that suitable training in health and safety has been given and experience gained in health and safety management.

Why manage occupational health?

Cost

There is a very strong business case for managing occupational health whatever the size of the company. In fact, the smaller the company, the greater the impact can be if employees fall ill.

The Health and Safety Executive (HSE) has published figures in *The Cost of Accidents at Work* which relate to different industries and include the construction industry. In 1990 some 750 000 employees suffering from ill health caused or exacerbated by their work had to take time off (*Labour Force Survey, 1990/91*). This resulted in the loss of over 30 million working days and a cost to industry of almost £700 million (Davies and Teasdale, 1994). These figures do not take into account the costs which fall outside of the employer on other sections of society such as National Health Service provision and social security, and the financial and non-financial losses to the victims suffering occupational ill health.

The Law

Morally, society does not accept ill health caused by work and as such has continually sought to address this by the provision and maintenance of a legal framework. Apart from the wider legislative requirements covering safety, the main pieces of legislation covering the management of occupational health hazards in the construction industry are:

- ☐ Control of Substances Hazardous to Health Regulations 1994

- Construction (Health, Safety and Welfare) Regulations 1996
- Construction (Design and Management) Regulations 1994
- Manual Handling Operations Regulations 1992
- Personal Protective Equipment at Work Regulations 1992
- Noise at Work Regulations 1989
- Control of Asbestos at Work Regulations 1987
- Control of Lead at Work Regulations 1980

It will not be possible within the confines of this chapter to look in detail at all of the above legislation. However, given that the Control of Substances Hazardous to Health (COSHH) Regulations 1994 is the principal piece of legislation for managing hazardous substances in the construction industry, the core of these Regulations will be discussed and applied in detail later as part of a management strategy for compliance. It should be noted that the COSHH Regulations do not apply to asbestos, lead and sources of ionizing radiation for which other, more specific Regulations and supporting Approved Codes of Practice apply.

Routes of entry for hazardous substances

Substances considered harmful to the body can enter by three primary routes—absorption, inhalation and ingestion:

- *Absorption* occurs mainly as a result of substances penetrating the skin and eyes *en route* to the blood stream. The effects may be localized (e.g. skin irritation) or may target organs elsewhere in the body (e.g. liver and kidneys). Examples of absorption toxins include chemicals such as solvents and microbiological agents such as leptospirosis.
- *Inhalation* of dusts, vapours, gases, fumes and mists can all result in ill health. This is probably the most common route of entry for hazards, and having entered the body in this way can be absorbed readily into the bloodstream. Solvent vapours, welding fumes, fibres and dusts are all inhalation toxins.
- *Ingestion* through the mouth is the least likely route of entry for hazardous substances. Smoking and poor hygiene arrangements often contribute to this route of entry. Lead and arsenic are well-known examples of ingestion toxins.

Occupational exposure limits

An important part of any occupational health management system is an assessment of the extent to which a hazardous substance can cause harm. This can be done simply by measuring physical and chemical factors in addition to exposure duration and applying to them occupational exposure limits contained within EH40/97, *Occupational Exposure Limits*. This is a publication produced annually by the HSE and contains an up-to-date list of occupational exposure limits for use with the COSHH Regulations.

EH40/97 contains two types of occupational exposure limits for hazardous substances, namely:

- OES—occupational exposure standard
- MEL—maximum exposure limit.

An *OES* is set at a level that (based on current scientific knowledge) will not damage the health of workers exposed to it by inhalation day after day.

An *MEL* is set for a substance which may cause the most serious health effects such as cancer and occupational asthma and for which safe levels of exposure cannot be determined, or for substances for which safe levels exist but control to those levels is not reasonably practicable.

Both types of limit are concentrations of hazardous substances in the air, averaged over a specified period of time referred to as a 'time-weighted average' (TWA). Two time periods are used, namely:

- long term—8 h
- short term—15 min.

Short-term exposure limits (STELs) are set to help prevent effects such as eye irritation, which may occur following exposure for a few minutes.

It is important to remember that occupational exposure limits such as OESs and MELs are fundamental in assessing the risk of harm to workers and must always be taken account of prior to the exposure of a chemical or other hazardous substance to a worker.

Key hazardous substances commonly found in construction

The accumulation of information on materials containing hazardous substances prior to their specification and use is an essential first step in the prevention and control of hazardous substances. There are, of course, many sources of information where useful health and hygiene data can be found. However, there exists widespread confusion among professionals with regard to terminology. Two of the most common are:

(a) types of data sheets
(b) technical terms found on these sheets.

With regard to (a) there are principally two types of data sheet found in construction, namely:

- Material data sheets—these are usually material performance sheets containing performance specifications.

- Safety data sheets—as the name suggests, these sheets are the main source of safety data and should be used when making COSHH assessments. These sheets incidentally are required to be produced by law when a chemical product is classified as dangerous and is supplied for use at work and should contain at least 16 sections with appropriate information therein.

With regard to (b) above, the other area of concern relates to technical terms. Examples of these include:

- LD_{50}
- NOAEL
- necrosis
- synergistic
- MEL
- OES
- TWA
- ppm
- mg/m^3

Although these terms have definitions that can be found in key HSE publications including EH40 and EH64, *Summary Criteria for Occupational Exposure Limits*, it is important to be able to assimilate their detail. This would be a perfect example of when assistance should be sought from other professionals as outlined earlier.

Contained in Appendix 1 in the form of three tables is a list of key hazardous substances and processes commonly found in construction, namely dust and fibres, fumes and gases, and chemical products. The tables contain some key control measures useful in the prevention and control of the substance. The control measures are generic in what would be required; however, a detailed risk assessment would be required based on more extensive safety data in order to comply with the COSHH Regulations.

Control of Substances Hazardous to Health Regulations 1994

General

The COSHH Regulations are the principal piece of UK legislation for managing occupational health and hygiene. In essence the Regulations:

- Place a responsibility on employers to do what is reasonably practicable to ensure the safety of employees by protecting them from harmful substances.
- Outline the essential requirements in a systematic way for controlling hazardous substances and for protecting people exposed to them.

Employers are required to:

(a) carry out an assessment of the risk to health created by work with a view to determining what precautions are needed (Regulation 6)
(b) introduce appropriate control measures to prevent or control the risk (Regulation 7)
(c) ensure the use of control measures (Regulation 8)
(d) ensure that whatever control measure is used, is maintained, examined and tested (Regulation 9)
(e) monitor exposure at the workplace (Regulation 10)
(f) carry out appropriate forms of health surveillance (Regulation 11)
(g) inform, instruct and train people about the risk and precautions to be taken (Regulation 12).

Before examining each of the above in more detail, note that the Health and Safety Commission (HSC) has published detailed guidance in the form of Approved Codes of Practice (ACOPs) on:

- control of substances hazardous to health
- control of carcinogenic substances
- control of biological agents.

The ACOPs have special legal status in that if a company is prosecuted for breach of health and safety law and it can be proved that the relevant provisions of the ACOP have not been followed, the company will be found at fault unless it can show that it has complied with the law in some other way. Each of the above ACOPs are formatted in the same way as (a)–(g) above and provide the key elements of a COSHH management system.

What is a substance hazardous to health?

A precise definition of a substance hazardous to health can be found in the COSHH ACOP (HSC publication ACOPs (L5)), and generally comprises:

- A substance which is listed in Part I of the approved supply list as dangerous for supply within the meaning of the Chemicals (Hazard Information and Packaging for Supply) Regulations 1994 and for which an indication of danger specified for the substance in Part V of that list is very toxic, toxic, harmful, corrosive, sensitizing or irritant.
- A substance listed in Schedule 1 of the COSHH ACOP which lists substances assigned a maximum exposure limit (MEL).
- A biological agent.
- Dust of any kind when present at a substantial concentration in air.
- A substance, not mentioned above, but creating a hazard to health comparable in nature to those mentioned above.

Competence

There are many references to 'competence' in health and safety law, one of which is contained in Regulation 6(5) of the Management of Health and Safety at Work Regulations 1992. It states that: 'A person shall be regarded as competent ... where he has sufficient training and experience or knowledge and other qualities'. The COSHH Regulations are no different in this respect in that the ACOP states that: 'any person who carries out any work on behalf of the employer in relation to any of his duties under the Regulations should possess sufficient knowledge, skill and experience to be able to perform that work effectively'.

The employer should therefore ensure that the person to whom any work is delegated is competent for that purpose. This may entail engaging additional expertise from within or outside the company. Some tasks under the Regulations may require a range of expertise that may not be possessed by a single person in the company. Therefore the employer may appoint one or more of their own employees to do all that is necessary, or enlist support from outside the organization, or do both. So, with the need to be competent in mind, let us now look in detail at the key elements that go to make up a COSHH management system.

Five steps fundamental to compliance with the COSHH Regulations

Figure 2 illustrates five key element/steps fundamental to compliance with the COSHH Regulations, and is based on the principles of risk

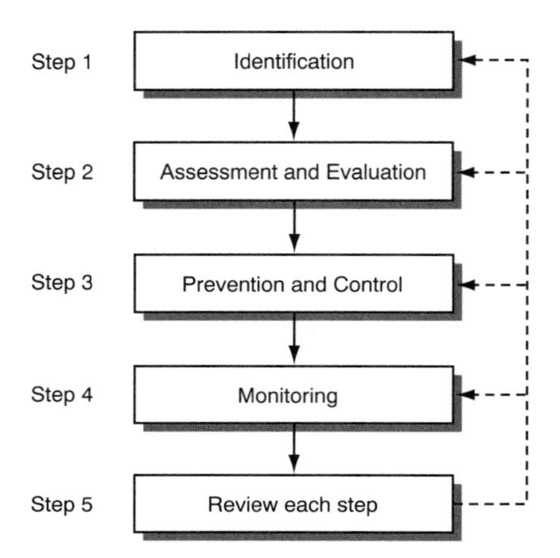

Fig. 2. Five steps fundamental to compliance with COSHH

management, that is identification, evaluation and control. The following five steps are based on Regulations 6–12 of the COSHH Regulations, and contain original information from the ACOP. This should ensure the accuracy of the guidance which is essential in complying with the COSHH Regulations.

Step 1: identify the hazards

The key to successfully identifying hazardous substances is to utilize one or more of the following techniques as an information-gathering exercise:

- ☐ Walk around the workplace and check for substances and processes that might cause harm.
- ☐ Keep trivia in perspective and concentrate only on significant hazards.
- ☐ Consult with the operatives involved with the hazardous substance or process.
- ☐ Gather information on the hazards found, for example manufacturers' instructions and data sheets (NB: manufacturers have a legal duty to provide this information).
- ☐ Look for hazardous substances that may be on site before work starts, for example sewer gases, contaminated landfill, microbiological agents (e.g. leptospirosis).
- ☐ Check with others such as the client, planning supervisor, designers, principal contractor, contractors and the health and safety professional for information and experience on managing hazardous substances.
- ☐ Check the health and safety plan.
- ☐ Seek advice from other professionals.

Step 2: assess and evaluate the risks arising from the hazards and decide whether existing precautions and controls are adequate or more should be done

First—decide who might be harmed and how

- ☐ Site operatives and office staff.
- ☐ Members of the public, for example passers-by and children.
- ☐ Visitors to site.
- ☐ Cleaners.
- ☐ Visiting contractors.
- ☐ Consider the interface between the people engaged in the process or substance and other site staff.
- ☐ Pay particular attention to those who are vulnerable—inexperienced staff, lone workers, people with disabilities, operators, etc.

Second—assess the risks

An assessment is required of the level of risk to the person or persons identified above. Consider the two main areas of concern, that is, the severity of the hazard and the likelihood of harm occurring. This should be based against any controls currently in place.

A simple matrix based on severity and likelihood could be used amongst others, to establish a risk rating for use in prioritizing action. See Fig. 3: to obtain a risk rating, simply assess the severity of hazard on the left-hand side of the matrix and go across horizontally stopping at where you believe the likelihood of harm occurring exists. Once a risk rating has been established, apply the information contained in the risk rating section as a means of prioritizing the actions.

Severity (S)

High — fatality, major injury or illness causing long-term disability (including carcinogenic substances and those with MEL)

Medium — injury or illness covering short-term disability (include substances with an OES)

Low — other injury or illness (substance not contained in EH40, *Occupational Exposure Limits*)

Likelihood (L)

High — certain or near death

Medium — reasonably likely

Low — unlikely

Risk rating

3 = high risk—additional preventative or control measures necessary to reduce risk rating to an acceptable level

2 = medium risk—as above unless good reason

1 = low risk—no further precautions or controls necessary at this stage but review when circumstances change

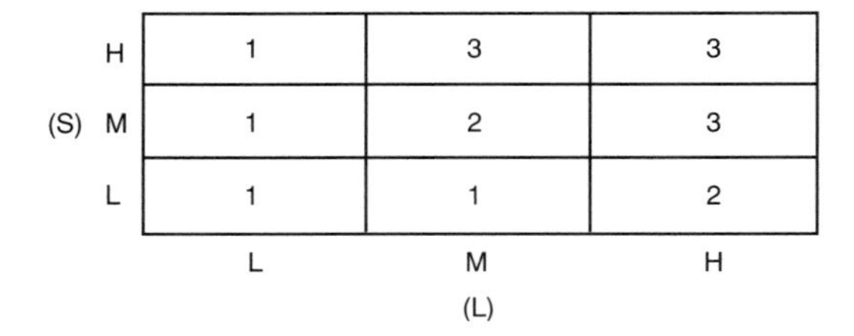

(S)	L	M	H
H	1	3	3
M	1	2	3
L	1	1	2
	(L)		

Fig. 3. Risk rating matrix for use in prioritizing actions

Third—ask the question

- ☐ Do any existing precautions or control measures meet current legal requirements?

NB: If the answer to the above is yes, then the risks are adequately controlled, but you will need to indicate in your risk assessment any controls that you may have in place. You may refer to procedures, company rules and manuals, etc. If the answer to the above is no, continue on to step 3.

Step 3: apply suitable prevention or control measures

Carcinogens

- ☐ Is the substance carcinogenic as listed in EH40, *Occupational Exposure Limits*?

If the answer is yes, continue on. If the answer is no, then go straight to 'non-carcinogens' and apply the control measures in points (h)–(r).

- ☐ If the substance is carcinogenic, then you must, as far as is reasonably practicable, prevent exposure to a carcinogen by using a safer alternative substance or process. Then return to the beginning of step 1 and repeat the process on the alternative/safer substance or process.
- ☐ If it is not reasonably practicable to prevent exposure to a carcinogenic substance by using an alternative substance or process, then all of the following control measures should be applied:

(a) Total enclosure of the process and handling system The use of totally enclosed systems should be the first choice of control measure for carcinogens. NB: If it is not reasonably practicable totally to enclose the process and handling system the following measures (b)–(g) are still required.

(b) The use of plant, processes and systems of work which minimize the generation of, or suppress and contain spills, leaks, dust, fumes and vapours of carcinogens Where (a) above cannot be achieved, plant processes and systems of work should be designed and operated so that they minimize the generation of, suppress and contain carcinogenic substances. Examples would include partial enclosure, handling systems, local exhaust ventilation (LEV) and general ventilation.

(c) Limit the quantities of a carcinogen at the place of work Carcinogenic substances intended for use should be kept to the minimum needed for the process and stored and transported on site in closed containers, clearly labelled and with clearly visible warning and hazard signs.

(d) Keep to a minimum of the number of persons who might be exposed to a carcinogen The areas in which exposure to carcinogens may occur should be clearly identified and measures taken to prevent the spread of contamination within and beyond these areas. The number of people in the area should be kept to its minimum with the duration of exposure being limited to what is necessary to be able to do the work.

(e) The prohibition of eating, drinking and smoking in areas that may be contaminated by carcinogens Eating, drinking and smoking should not be allowed where there is a risk of contamination from a carcinogenic substance. Suitable warning signs should be used and dedicated areas provided where people can eat, drink and smoke. Adequate washing facilities in conjunction with suitable hygiene measures should be provided.

(f) The designation of those areas and installations which may be contaminated by carcinogens and the use of suitable and sufficient warning signs Suitable signs in conjunction with a segregated area should be used to distinguish between clean and contaminated areas. If possible physical protection should be used to restrict the chance of accidental intrusion to the contaminated areas.

(g) The safe storage, handling and disposal of carcinogens and use of closed and clearly labelled containers Apart from the risks associated with the 'use' of a carcinogen it is equally important to ensure the safe storage, handling and disposal of it. Sealed containers, handling systems and environmental controls in conjunction with any personal protective equipment (PPE) should all be adopted.

Non-carcinogens

For hazardous substances not classified as carcinogens, where prevention of exposure is not reasonably practicable, adequate control of exposure should be achieved by measures other than PPE. This may include any combination of the following:

(h) totally enclosed process and handling systems
(i) plant, processes or systems of work which minimize generation of, or suppress or contain the hazardous dust, fume, biological agent, etc., and which limit the area of contamination in the event of spills and leaks
(j) partial enclosure with LEV
(k) LEV
(l) sufficient general ventilation
(m) reduction of numbers of employees exposed and inclusion of non-essential access
(n) reduction in the period of exposure for employees
(o) regular cleaning of contaminant from walls, surfaces and work areas

(p) provision of means for safe storage and disposal of substances hazardous to health
(q) prohibition of eating, drinking, smoking, etc., in contaminated areas
(r) provision of adequate facilities for washing, changing and storage of clothing and PPE.

PPE

Only as a last resort and having shown that exposure to a hazardous substance could not have been prevented or adequately controlled, can suitable PPE be used. This will be on the basis that the PPE will adequately control exposure to those substances (see the HSE publication *Personal Protective Equipment at Work Regulations 1992*, Schedule 1).

Examples of situations where the use of suitable PPE may be necessary include:

- ☐ where it is not technically feasible to achieve adequate control of exposure by process, operational and engineering measures alone
- ☐ as a temporary measure until adequate control is achieved by other means
- ☐ in emergency situations where it is the only practicable solution
- ☐ during routine maintenance operations

Information, instruction and training for persons who may be exposed to substances hazardous to health

A key requirement of the COSHH Regulations is to provide suitable and sufficient information, instruction and training to all people at risk from hazardous substances. It could be argued that this requirement is the most important part of any set of controls, particularly where people are involved and play such an important part in applying them.

The *information* provided to employees and others who may be affected should include in particular:

- ☐ the nature and degree of the risks to health arising as a consequence of exposure including any factors that may increase the risks such as smoking
- ☐ the reason for any control measures to be adopted and how to use them properly
- ☐ the reasons for using any PPE including where and when they are necessary
- ☐ any monitoring arrangements including access to results, etc.
- ☐ the role of health surveillance and the arrangements for access to individual health records and the collective results of health surveillance.

Any *instruction* to people must be suitable and sufficient for them to know what they should do, the precautions to take and when they should take them. It should include any cleaning, storage and disposal procedures, the reason why they are required and when they are to be carried out.

All *training* should be to a suitable level to ensure that persons at work can effectively apply and use the methods of control, PPE and emergency measures.

Step 4: monitoring

Adequate control of exposure

As with any management system, there is a need to monitor its effectiveness. Managing the COSHH Regulations is no different in that the need to monitor exposure of substances hazardous to health is not just good practice but a legal requirement.

The control of exposure to any substance assigned an MEL shall only be treated as adequate if the exposure is reduced so far as is reasonably practicable and in any case below the MEL. This level should *never* be exceeded.

The control of exposure to a substance assigned an OES shall only be treated as adequate if the exposure is reduced to or below the OES. The difference between an OES and an MEL in relation to adequately controlling exposure is that if the exposure by inhalation exceeds the OES, then control will still be deemed to be adequate provided that the employer has identified why the OES has been exceeded and is taking appropriate steps to comply with the OES as soon as is reasonably practicable.

Measuring levels of exposure

Compliance with occupational exposure limits may be demonstrated by measuring and recording the exposure of employees in accordance with the principles contained in HSE Guidance Note EH42, *Monitoring Strategies for Toxic Substances*. This document provides practical guidance on monitoring substances hazardous to health in air. For some substances, biological monitoring may also be appropriate. Biological monitoring is the measurement and assessment of hazardous substances or their metabolites in tissues, secretions, excreta or expired air. It can be a very useful complementary technique to air monitoring when air-sampling techniques alone may not give a reliable indication of exposure.

The selection of a particular monitoring strategy will require specialist advice and will ultimately be subjective as it depends on a variety of considerations. There are however, specific substances and processes for which monitoring is a statutory requirement with predetermined frequencies. This information can be found in Schedule 4 of the COSHH ACOP.

Records

It is important to remember that whatever monitoring strategy is selected, suitable records should be kept of the monitoring carried out. Typically a suitable record should provide sufficient information to determine:

- ☐ when the monitoring was done
- ☐ what the results were
- ☐ what monitoring procedures were adopted, including duration
- ☐ locations where samples were taken
- ☐ operations in progress at the time
- ☐ names and details of individuals concerned.

Records should be kept available for inspection for at least five years unless they relate to particular individuals where the period is for at least 40 years.

Health surveillance

The objectives of health surveillance in relation to occupational exposure to hazardous substances and processes are:

- ☐ to protect employees' health by early detection of adverse changes due to an exposure of a hazardous substance
- ☐ to assist in the evaluation of the measures taken to control exposure
- ☐ the collection, maintenance and use of data for the detection and evaluation of hazards to health
- ☐ to assess, in relation to specific work activities involving biological agents, the immunity of employees.

Health surveillance should be treated as being *appropriate* where:

- ☐ an employee is exposed to one of the substances specified in Schedule 5 of the COSHH ACOP, or
- ☐ exposure of a substance hazardous to health is such that an identifiable disease or adverse health effect may be related to the exposure and could manifest itself under working conditions. This is providing that there are valid techniques for detecting indications of the disease or the effect.

Further details on health surveillance can be found in the COSHH ACOP.

Step 5: review

The HSE, in *Successful Health and Safety Management*, defines reviewing as: 'the process concerned with making judgements about the adequacy of performance and taking decisions about the nature and timing of the actions necessary to remedy deficiencies'.

The final step in the process of managing hazardous substances is the requirement to review the risk assessment. The assessment should be reviewed regularly and in any case where there is any evidence to suspect that it is no longer valid or where there has been significant change in the work or process to which the assessment relates.

The assessment may be *no longer valid* because of, for example:

- the results of periodic thorough examination and tests of engineering controls
- the results of monitoring exposure at the workplace
- the results of monitoring or a confirmed cause of an occupationally induced disease
- new information on health risks.

A *significant change* in the work may be:

- in the substances used or their source
- plant modification, including engineering controls
- in the process or methods of work
- in the volume or rate of production

The assessment should include the date on which the assessment is made, followed by the maximum period which should elapse before successive reviews. The length of the period chosen will depend on the nature of the risk, the work and a judgement on the likelihood of changes occurring, but in any case the assessment should be reviewed at least every five years.

Summary

Managing hazardous substances in construction requires knowledge, skill and experience in order to apply the legislative requirements correctly. To carry out this process efficiently requires a logical and structured approach comprising five key steps:

Step 1 identification
Step 2 assessment and evaluation
Step 3 prevention and control
Step 4 monitoring
Step 5 review

In carrying out these five steps you may need to seek advice from health, safety and hygiene professionals along the way, and remember that the provision of suitable information, instruction and training to all those using the substance and affected by it is arguably the most important part of the process.

Appendix 1: key hazardous substances and processes commonly found in construction

Table 1. Dusts and fibres

Hazardous substances	Tasks	Key control methods
Cement: dermatitis from chromate impurities, skin burns, harmful to mouth and nose from lime content and when wet	Masonry and plaster work in particular Concrete	Minimize spread of material, respiratory protection for dry mixing/ handling, gloves, waterproof boots, personal hygiene, barrier creams before and after working
Manmade mineral fibre: rockwool; irritant to respiratory tract, eyes and skin	Insulation work in particular Bricklaying and carpentry in cavities and walls	Minimize cutting and handling, respiratory protection when MEL likely to be exceeded. One-piece overall, gloves
Gypsum: irritant to throat, nose and eyes	Plastering and masonry work	Control—see cement
Silica: silicosis, and increased risk of respiratory complaints	Grit blasting of masonry, concrete scabbling, granite polishing, tunnelling in silicate rock, power cutting of furnace brickwork/liners	Wet methods, process enclosure with dust extraction; respiratory protection
Wood-dust: irritants, allergic reactions (e.g. western red cedar, and other hardwoods). Resin bonded materials very irritating	Carpentry work. Most problems connected with use of power machinery, e.g. belt sanders	Off-site preparation On-site: enclosure and exhaust ventilation, dust extraction on portable tools

Source: Construction Health and Safety Manual.

Table 2. Fumes and gases

Hazardous substances/ processes	Tasks	Key control methods
Welding, brazing, cutting produce a wide variety of fumes depending on metals being worked on, the electrodes used, fluxes, etc. Fumes are highly irritating to respiratory system (chronic mainly in effect). Main gases evolved are carbon monoxide, nitrous fumes and ozone	Welding, brazing and cutting. Confined spaces are particularly hazardous	Local exhaust ventilation first choice for confined spaces; good general ventilation; air supplied helmet. Monitoring of exposure
Hydrogen sulphide: extremely toxic; irritates eyes, nose and throat and potentially lethal	Work involving sewers, drains, excavations in made ground, demolition of sulphur-trapping plants in refineries	Confined space procedures; exhaust and forced ventilation; airline/self-contained apparatus: monitoring
Carbon dioxide: an asphyxiant heavier than air	Bore holes in chalk and limestone, CO_2 welding in confined spaces	See above
Carbon monoxide: toxic	Operation of LPG equipment, petrol or diesel plant in or close to confined spaces	Site away from confined spaces; mechanical ventilations: exhaust filter efficiency

Source: Construction Health and Safety Manual.

Table 3. Chemical products

Hazardous substances	Tasks	Key control methods
Solvents		
Toleune, xylene, 1,1,1-trichloroethane etc.—are present in many construction products, e.g. paints, lacquers, glues, strippers, as thinners. Solvents are harmful through inhalation of fume (or accidental ingestion) and via skin contact—dermatitis can result	Used in many activities, but particularly decorative applications, tile fixing, use of resin systems on site. With most materials, risks increase in relation to quantity used and frequency/duration, particularly spray application or work in ill-ventilated/confined spaces	Select safest materials and method of application. Ensure good ventilation always. Confined spaces require mechanical ventilation/use of airline or self-contained breathing equipment, and similar standards for spray work. 'Airless' or 'mist-less' spray techniques should be considered. Impervious protective clothing and good washing facilities/barrier creams are important
Resin systems		
Isocyanates, e.g. MDI, TDI, polyurethane surface coatings or adhesives. Known respiratory irritant causing asthma and sensitization. Paints are hazard when brushed or rolled	MDI for thermal insulation of buildings (e.g/ roof spraying) Polyurethane for decorative work by brush, roller or spraying; also one- and two-pack coatings	REP; impermeable protective clothing. Applications in confined spaces—mechanical ventilation, breathing apparatus and protective clothing Brush roller applications in normal conditions—good normal ventilation for spraying—breathing apparatus offers best protection. Isocyanates: segregate mixing areas; provide good ventilation and personal protective equipment, washing facilities
Epoxy: severe irritant and sensitizer; toxic, particularly to liver	Work using high-strength adhesives for joining structure units, floor tube and pipe coatings	
Polyester: styrene vapour both toxic by inhalation (liver), also narcotic and irritant to eyes and skins	Glass fibre-reinforced structure work, claddings and coatings	Good ventilation; mechanical ventilation where practicable, eye protection and RPE, washing facilities. Again, confined workspace requires high standards of control

Table 3. Continued

Hazardous substances	Tasks	Key control methods
Preservatives/fungicides		
Fume, irritation of skin, damage to nervous system and other organs from range of active ingredients	*In situ* timber treatment, particularly confined space work of long duration	See HSE Guidance Note GS46
Lubricants		
Mineral oils cause dermatitis, acne and possibly skin cancer in extreme cases; respiratory damage in mist form	Near machinery, mould release agents: formwork mist from compressors and air tools in confined spaces	Filtration to reduce mist, good ventilation, respiratory protection and protective clothing (impervious to oil) personal hygiene
Acids/alkalis		
Hydrochloric, hydrofluoric and sulphuric acids commonly used. Corrosive action on skin if in contact with stonework, etc.; fume causes respiratory irritation	Chiefly masonry cleaning	Use weakest concentrations possible. Skin and eye protection: personal hygiene
Site contaminants		
These hazards are present in the soil/structures arising from previous 'industrial' activities, or exist naturally: toxic metals/materials at gasworks, tanneries, hospitals, e.g. cadmium, arsenates, cyanides, phenols; dangerous by inhalation, ingestion and skin contact/absorption. Microbiological risks include Weil's disease, tetanus, hepatitis B	Site redevelopment involving groundwork, demolition, tunnelling activities in particular, work near contaminated water courses	Thorough site examination and clearance. Respiratory protection and protective clothing. Highest hygiene standards. Immunization against tetanus

Source: Construction Health and Safety Manual.

Bibliography

Construction Industry Publications (1996). *Construction Health and Safety Manual 1996*. Construction Industry Publications, Birmingham.

Davies, M.V. and Teasdale, P. (1994). *The Costs to the British Economy of Work Accidents and Work-related Ill Health*. HSE, Sudbury.

Health and Safety Commission (1997). *Control of Substances Hazardous to Health Regulations 1994. Approved Code of Practice*. HSE, Sudbury, L5.

Health and Safety Executive (1989). *Monitoring Strategies for Toxic Substances*. HSE, Sudbury.

Health and Safety Executive (1990). *Health Surveillance under COSHH: Guidance for Employers*. HSE, Sudbury.

Health and Safety Executive (1990). *Assessment of Exposure to Fume from Welding and Allied Processes*. HSE, Sudbury, EH54.

Health and Safety Executive (1989). *COSHH: An Open Learning Course*. HSE, Sudbury.

Health and Safety Executive (1990). *Control of Exposure to Fume from Welding, Brazing and Similar Processes*. HSE, Sudbury, EH55.

Health and Safety Executive (1992). *Biological Monitoring for Chemical Exposures in the Workplace*. HSE, Sudbury, EH56.

Health and Safety Executive (1992). *Management of Health and Safety at Work Regulations 1992*. HSE, Sudbury.

Health and Safety Executive (1992). *Personal Protective Equipment at Work Regulations 1992*. HSE, Sudbury.

Health and Safety Executive (1993). *A Step-by-step Guide to COSHH Assessment*. HSE, Sudbury.

Health and Safety Executive (1993). *Successful Health and Safety Management*. HSE, Sudbury, HS(G)65.

Health and Safety Executive (1994). *Preventing Asthma at Work: How to Control Respiratory Sensitisers*. HSE, Sudbury, L55.

Health and Safety Executive (1996). *Summary Criteria for Occupational Exposure Limits*. HSE, Sudbury, EH64.

Health and Safety Executive (1996). *Chemicals (Hazard Information and Packaging for Supply) Regulations 1994 (SI 1994/3247) as amended by the Chemicals (Hazard Information and Packaging for Supply) (Amendment) Regulations 1996*. HSE, Sudbury, SI 1996/1092.

Health and Safety Executive (1997). *The Costs of Accidents at Work*, 2nd edn. HSE, Sudbury.

Health and Safety Executive (1997). *Occupational Exposure Limits*. HSE, Sudbury, EH40/97.

OPCS (1992). *The 1990/91 Labour Force Survey*. HMSO, London.

9. Subcontract management

Michael H. Evans, *John Laing plc, Hemel Hempstead*

Introduction

Within the construction industry there are a variety of contractual arrangements and statutory obligations relating to the management of the subcontracting process. For consistency of approach and simplicity of explanation this chapter is based upon the scenario of a principal contractor looking to subcontract some of his work. The areas of planning, choosing subcontractors, working with subcontractors, monitoring and review are covered, and both the theory and practice of each stage are included.

Subcontracting has long been an essential part of the construction process and the use of subcontractors has increased steadily over recent years. Traditionally it was seen simply as a way of buying in specific expertise or specialist services. Today, however, subcontractors are used for a variety of other reasons as well: to off-load non-core business, to spread commercial risk, to offer clients total package services which go beyond a contractor's core business, and to better cater for peaks and troughs in workload.

The increased use of subcontractors and the decrease in direct employment have been described by some as a fragmentation of the industry and blamed for the continuing poor safety record in construction. Indeed, some major clients of the construction sector are advocating that the industry return to direct employment for its core activities and rely on subcontractors only for specialist services.

However, the use of subcontractors also has advantages. They tend to be more specialized and should therefore be more knowledgeable about their particular area of operation. The overall approach to a package of work can be more flexible and the lead in time from decision to action can be much shorter. With smaller subcontractors, lines of communication are also shorter, as not as many levels of management are involved.

Whatever trends may follow, it is certain that subcontracting will remain an important part of the construction process. It is therefore essential that subcontract management be recognized as a key issue for the industry, with health and safety being given equal consideration to all other business issues.

Planning

The decision to subcontract should not be taken as a last minute 'easy option'. It should be a strategic decision, taken at the tender stage, as part of the overall approach towards the project. The precise scope of subcontract packages needs careful consideration. Size, complexity, likely duration and interfaces all have to be considered against the project requirements, in-house resources and availability, as well as likely subcontractors. This is part of the overall planning of the construction phase.

The contractual issues are also important at this stage as they can set the scene for the sort of relationships that follow. Proposed contracts with subcontractors should be compatible and consistent with the main contract. Both should allocate the rights and obligations of the parties involved and the commercial risks to be borne by each. Both should also support the statutory requirements under the Construction (Design and Management) (CDM) Regulations 1994 and, where necessary, clarify precisely who is to do what in terms of health and safety requirements. However, it is vital that all parties recognize that they cannot contract out their statutory duties. *Statute always takes precedent over contract.*

In determining the scope of subcontract packages an initial risk assessment should be carried out, if these risks are not already apparent from the planning supervisor's pre-tender health and safety plan. As described elsewhere in this publication, this will involve identifying the major hazards and assessing the major risks arising from both the work process itself and the likely interfaces when the work is to be carried out.

The purpose of this assessment is twofold: first, it may enable risk elimination or reduction through varying the sequencing of operations; and secondly, the significant findings can be incorporated into the developing health and safety plan which can be included in the subcontract documentation. Depending on the particular circumstances and timing, this may either enable the subcontractor to have an input into further developing a particular aspect of the health and safety plan or at least enable him to make allowance in his pricing for the predetermined issues in the health and safety plan.

Anticipated input to a construction phase health and safety plan from a subcontractor will vary according to the size, nature and complexity of the subcontracted works and the stage of development of the plan when the package is placed. However, the guiding principle is that the subcontractor should be provided with the most up to date version of the health and safety plan as part of his tender documentation. The sometimes used technique of 'making documents available for the inspection of tenderers' is not recommended for any safety-related documentation. It must be made absolutely clear to the subcontractor at the tender stage what arrangements have been made, and what facilities are to be provided by the principal contractor or others and what the subcontractor is expected to contribute himself. This would include items such as site rules, welfare and first aid provisions, emergency arrangements, and any attendances such as craneage, access scaffolding, power and lighting.

In all dealings with subcontractors, and particularly with health and safety issues, the objective must be to enter into fair and transparent agreements which are aimed at improving the effectiveness and efficiency of the delivered product.

Choosing subcontractors

Subcontractors can be selected by various means: competitive tendering, negotiation or as a result of partnering or joint venture. Competitive tendering is still the most common method although the other arrangements are growing in popularity.

Competitive tendering is a complex process and requires everyone involved to follow a common set of procedures. This is as important for health and safety issues as for all other business issues. The Construction Industry Board publication *Code of Practice for the Selection of Subcontractors* sets out the selection process thus as

Qualification
↓
Compilation of tender list
↓
Tender invitation and submission
↓
Tender assessment
↓
Tender acceptance

and identifies a number of key principles in selecting subcontractors, including

- tender lists should be compiled systematically from a number of qualified candidates
- sufficient time and information should be provided to allow preparation of tenders appropriate to the type of works
- tenders should be assessed and accepted having regard to quality as well as price
- proposed contracts should be compatible and consistent with the main contract
- there should be a commitment to teamwork from all parties.

While health and safety is not specifically mentioned in any of the key principles identified, there is an obvious benefit if those identified above are achieved.

One of the keystones of the CDM Regulations is the issue of competence and resource. Under the regulations there is a statutory duty on anyone arranging for someone else to carry out construction work, to check the competence and resources of those that they select.

Factors to be considered in choosing a subcontractor generally include:

- ☐ suitability—does the nature and scope of the works fall within their general area of operation?
- ☐ availability—do they wish to bid for the work?
- ☐ cost—is their price reasonable?
- ☐ technical competence—do they have the necessary technical competence?
- ☐ reliability—what experience do we have of how they perform?

and must by law now include:

- ☐ health and safety—are they sufficiently competent and resourced to comply with all relevant legislation during the execution of their work?

It is becoming increasingly common that the assessment of both principal contractors and subcontractors is done as a two stage process:

(a) a general assessment as part of a pre-qualification process
(b) tender assessment, where the responses to project specific requirements are evaluated.

Prequalification assessments are used to prepare a list of potential tenderers, either for a one-off project or for a pool of 'approved' tenderers for various projects over a period of time. In the latter case the assessment may need to be updated from time to time. The assessment process should include:

- ☐ evaluation of the organization's previous experience of this sub-contractor
- ☐ assessment of their general reputation within the industry using contacts in other businesses and informal networks
- ☐ their response to specific questions posed at interview or via questionnaires.

Health and safety issues must be included as an integral part of each of these stages.

It is unfortunate that in the industry's rush to demonstrate compliance with the CDM Regulations, the popularity, length and complexity of prequalification questionnaires have grown out of all proportion to their usefulness. It is the opinion of the author that the value of questionnaires in assessing competence in health and safety is extremely limited. Organizations have generally become more adept at providing the 'right' answers, and there is frequently a large reality gap between the theoretical answers provided and the actual practices on site. Data on accident statistics and enforcement action are notoriously unreliable and cannot be used in isolation for comparative purposes. Is the small back-street building firm who has had no reportable accidents or enforcement action more competent in health and safety issues than the large national contractor who declares several of both? What are the criteria to be applied in making this judgement?

Assessment of health and safety issues is far more important at the tender evaluation stage than at the prequalification stage. It is also likely to be a far more realistic assessment at the tender stage since it can include evaluation of responses to project specific safety questioning on issues such as:

- contribution to or response to the construction phase health and safety plan
- method statements for specific operations
- key personnel
- general supervisory arrangements
- provision of training
- any sub-subcontracting
- health and safety monitoring arrangements and resources.

Prior to formally placing the subcontract it is often necessary, and in some companies compulsory, to interview one or more of the most preferred tenderers in order to clarify or amplify their submissions. This provides an ideal opportunity for confirmation of their real understanding of the health and safety issues, especially where high risk activities are involved. Any matters agreed at such interviews should be recorded in writing.

For major subcontractors, or high-risk work or in pursuance of long-term partnering arrangements with subcontractors, it is strongly recommended that similar work currently being undertaken by the tenderers be viewed. In this way an assessment can be made of their actual performance standards.

Working with subcontractors

Method statements

Once the subcontract has been awarded, but prior to the work commencing on site, a pre-start meeting should be held with the subcontractor. In many organizations this meeting facilitates the handover of the prime responsibility for the subcontract from the procurement department to the site management team. This is not to say that site management should not have been involved before this meeting—the greater their involvement, at as early a stage as possible, the better.

A significant element of this pre-start meeting should be devoted to health and safety, using the construction phase health and safety plan as the basis for the agenda. The attitude and expectations of the principal contractor at this meeting are crucial in determining how much effort the subcontractor will devote to health and safety issues. At this meeting all relevant health and safety information to date should be reviewed. This would include any comments from the planning supervisor's pre-tender health and safety plan, the current edition of the construction phase health and safety plan, the information provided to the subcontractor at the tender stage and the specific responses. The purpose of this review is to ensure that all parties share a common understanding of the precise health and safety requirements of the work, that it is based on the most up to-date information, including any interface problems, and that the proposed methods are acceptable to all concerned.

Method statements have long been used within the construction industry as a means of describing the manner in which elements of work are to be undertaken. Whilst historically they may have simply described a sequence of events, it is now commonly accepted that all method statements must also address the health and safety issues involved in carrying out the work. However, much of the work carried out within the industry is of a standard nature and, although it must of course be carried out safely, job specific method statements are not necessarily required for basic, routine operations. The broad range of risks posed by such operations are commonly addressed in the formulation of the project health and safety plan or in standard company procedures.

There is no definitive standard within the industry as to which operations need to be covered by a method statement or what that method statement should contain. It is recommended that, as a minimum, method statements should be prepared for all high-risk operations and any other operations where the control measures are not immediately obvious from the health and safety plan or other standard procedures. It is becoming increasingly common to require all subcontractors to produce method statements for all of their work. However, generalized statements of intent, dealing with large packages of work, as are often submitted with the tender, frequently prove to be of little use when it comes to carrying out the work.

Method statements should be written for the benefit of those carrying out the work and their immediate supervisors. They therefore need to address specific tasks or jobs, which may mean breaking down the subcontracted works into its constituent elements. Such job-specific method statements should address all aspects of the job—programme, sequence, engineering, method, quality etc., as well as health and safety.

They should be clear and concise, using simple sketches where appropriate, with no ambiguities or generalizations which could lead to confusion. Whilst the content will vary according to the scale and complexity of the job and level of risk involved, competent method statements should address the following issues:

- what is to be done?—the precise scope of the job

- ☐ where is it to be done?—the precise location
- ☐ when is it to be done?—by dates or by sequence of events, that is, preceding/concurrent with/following operations
- ☐ who is to do it?—the number and type of personnel including any specific skills, training or qualifications required
- ☐ how is it to be done?
  - the plant, equipment and materials required, including access, storage and handling thereof
  - a safe means of access and egress
  - a safe place of work
  - the precise method and sequence of operations, including any 'hold' points
  - the specific risks involved and how they are to be removed, minimized or controlled
  - the personal protective equipment and other safety equipment
  - the need for temporary works, and who is to design, supply and install
  - the specific limitations or constraints upon the job, for example adverse weather, out-of-sequence working
- ☐ references—the method statements should be self sufficient but it is occasionally necessary to cross-refer to specific drawings, specifications, standard procedures, etc.
- ☐ date and originator.

The following system controls also need to be specified, either as part of the method statement or as part of the project health and safety plan or company procedures.

- ☐ communications—arrangements for ensuring that all involved understand the method statement or their part of it
- ☐ supervision—arrangements for ensuring that work proceeds according to the method statement
- ☐ amendments—arrangements for agreeing modifications to the method statement and communicating them to those concerned
- ☐ validation—arrangements for ensuring that the proposed method statement is reviewed by both the subcontractor undertaking the work and the principal contractor, including the seeking of advice from the respective safety advisers as may be appropriate.

The Health and Safety Executive has made it clear that it considers method statements to be an integral part of the construction phase health and safety plan and therefore within the responsibility of the principal contractor under CDM Regulation 15(4). Successful prosecutions have already been brought against principal contractors because of the inadequacy of subcontractors' method statements.

Competent method statements, properly implemented, will eliminate the ad-hoc methods of work and good intentioned improvization which are so often the cause of accidents, substandard quality and poor productivity. It is the opinion of the author that job-specific method statements are by far the most important element in planning for safety and working safely.

Induction training

All persons working on construction sites should receive some form of induction course before they start work on any site. This introduction to the site should be aimed at providing everyone with the information they need to enjoy a healthy, safe and mutually beneficial period of employment at this location. While such courses tend to be dominated by health and safety issues, information on a whole range of other issues can also be included. For example: a brief description of the project; the facilities available to the workforce; industrial relations; quality and environmental issues; protecting the public; and being a good neighbour. Inductions are important for everyone on site, but particularly so for subcontractors as they are less likely to be familiar with many of the issues than the principal contractor's direct employees.

In health and safety terms, a good induction course can go a long way to help satisfy the requirements of the CDM Regulations upon the principal contractor to ensure that all persons on site have information about the risks and the site rules which apply. The specific health and safety content of an induction course will vary according to the nature of the work, the level of risk and the other arrangements for training and for provision of information. However, a typical course should cover the following points:

- a brief description of the project and the various parties involved—including names of principal contractor and planning supervisor
- a reminder of the employer's and employees' legal responsibilities for health and safety
- particular hazards on the project
- the site rules
- details of how health and safety will be managed on the project, especially means of communication, consultation and feedback such as safety committees, safety briefings and toolbox talks
- specific safety standards, procedures or permits, including training requirements
- use of method statements
- a safety monitoring regime
- emergency procedures
- first aid and welfare facilities

- ☐ health issues
- ☐ accident, incident and near-miss reporting
- ☐ security arrangements
- ☐ protection of the public.

It is recommended that a short and relatively simple questionnaire be used at the end of the induction course in order to encourage attention during the course and to verify that a certain level of understanding has been achieved by the attendees.

Records of those attending should be maintained, together with a list of the topics covered. On sites where security passes are required to gain entry, the issue of such passes can be linked to attendance at the induction course. Helmet stickers or entries on personal training record cards may also be used as a means of checking attendance.

The induction course should generally be designed and delivered by a representative of the principal contractor. Delegating it down the line and making subcontractors responsible for delivering their own inductions is not recommended. Similarly, the more senior and authoritative the person delivering the induction the more effective it is likely to be.

Whilst greater use is now made of induction videos, handout booklets and lists of site rules, the induction process should not rely entirely on such items. Induction courses afford an excellent opportunity for site management to begin a dialogue with the workforce and to clearly set the standards they wish to apply to their project. The benefits achieved will be proportional to the effort expended.

Control measures

Establishing and maintaining control is central to all management functions, including health and safety. Having entrusted an element of work to a subcontractor, the principal contractor cannot turn his back on it and assume that the subcontractor will adequately control that work. However, the principal contractor should not have to supervise the work directly; that is an integral part of the subcontractor's job. The principal contractor's role is to check that the subcontractor has the necessary controls in place.

A key issue in the subcontractor's control system will be the level of supervision he provides; this should have been agreed with the principal contractor prior to starting on site. The actual level will depend upon the complexity of the work, the risks involved and the competence of those carrying out the work to deal with those risks. However, even the most competent individuals should have some supervision, in order to ensure that the required standards are being achieved. The required competence of the supervisors themselves will also vary according to similar factors and also according to the level of authority and autonomy granted to the supervisors.

Good method statements are vital tools for supervisors, whose responsibilities include the passing of information to the individuals carrying out the work. This may be verbally, through briefing sessions or toolbox talks, or by copy of the actual method statement. The means of communication is of secondary importance to the objective—ensuring that everyone understands exactly what is expected of them. The next key role of the supervisor is to ensure that all the items required for the job are brought together at the right time, in the right place—the labour, plant, equipment, materials and the specified safety precautions. Then the supervisor must ensure that the work is indeed carried out as per the method statement. If, for any reason, this is not possible, the supervisor must ensure that the method statement is changed and duly approved. Short term expediencies and improvizations are frequently part of the cause of accidents.

For all but the smallest and simplest of subcontracts, changes to the scope or programme of works are all too common and detailed planning or replanning is an ongoing requirement. This should involve a reassessment of the risks posed, especially from changed interfaces or sequences of work, and the production and approval of new method statements as appropriate. Planning is an ongoing process in an industry notorious for changing requirements.

Co-ordination and co-operation are key duties of the principal contractor under the CDM Regulations. The mechanisms by which these are achieved, in relation to subcontractors, often include the appointment of subcontract co-ordinators or package managers within the principal contractor's team. The intent is generally to have a nominated single point of contact for each subcontractor to ensure clear and effective channels of communication.

Many of the controls affecting the subcontractor's work may be imposed by the principal contractor or by the client. Permits are a classic example. To ensure uniformity of approach across a site, a single permit system may be imposed on everyone working on the site. The disadvantage of this approach is that the system is unfamiliar to the subcontractor and often more complex than he is used to, because it was designed to cover numerous types of operations. This frequently results in a poor understanding of the system and hence poor compliance. Whenever possible, systems should be tailored to suit the specific needs of a job and even of a subcontractor. Better still is to allow the subcontractor to use his own systems wherever possible—provided, of course, that their adequacy is checked by the principal contractor.

In assessing the adequacy of the subcontractor's control mechanisms it must be remembered that while the contractual arrangements may place responsibility for failure on the subcontractor the same is not true of health and safety legislation. Health and safety is a shared risk, and increasingly principal contractors are being held accountable for the actions or inactions of their subcontractors.

Communication

During a subcontractor's time on site, mechanisms must be adopted which ensure that key information can easily be passed back and forth between the principal contractor and subcontractor and effectively disseminated throughout the subcontractor's workforce. Such mechanisms may be both formal and informal, and may include the written word, face to face discussion and visible behaviour which sets an example.

In terms of controlling subcontractors, the most important documents are the health and safety plan and job-specific method statements. The former lets everyone know the framework for health and safety and the latter provides the precise details as to how tasks are to be undertaken safely. It is vital that everyone knows exactly what is expected of them and method statements are the prime tools in achieving this. It is obviously not necessary for copies of method statements to be issued to everyone, but they should certainly be issued to foremen or those directly supervising the task. How the information is then passed to those undertaking the task will vary according to the circumstances but may include verbal briefings, toolbox talks, a copy of some or all of the method statement or the issue of work instruction sheets.

Other forms of written communication on health and safety include notices, posters and newsletters. These need to be used selectively to target specific issues and to complement other efforts as part of a co-ordinated campaign to improve knowledge of particular risks. While some issues might be specific to a particular subcontractor, it should be the principal contractor who initiates and co-ordinates such campaigns.

Face-to-face discussions should allow an open and honest exchange of views and enable individuals to ask questions and make a personal contribution. Whilst subcontractors should have their own systems for safety meetings, briefings or toolbox talks, these can often be augmented by input from the principal contractor. That is not to say that the principal contractor should take over responsibility for these issues; his role should be to provide appropriate information, especially on site-wide issues, for the subcontractor's supervisors to disseminate further.

The written and verbal messages on health and safety must be supported by the visible behaviour of managers at all levels, who must lead by example. Any inconsistency between what is said and what is done, or seen to be condoned, will rapidly undermine all efforts at improvement of health and safety standards. Managers from both the principal contractor and the subcontractor organizations can signal their commitment to health and safety by methods such as:

- regular health and safety tours of the site incorporating random questioning of individual workers and supervisors on safety issues
- management involvement in induction and training sessions perhaps by providing the introduction
- a safety committee chaired by the most senior site manager
- managers becoming involved in accident and incident investigations.

It is very easy for senior managers to assume that everyone shares their commitment to health and safety and shares a common understanding of the issues involved. The perception of those lower down the organization as to what senior managers want of them is frequently at odds with this. Senior managers might well ponder why this should be! The imbalance must be addressed through both words and deeds.

Co-operation

The Management of Health and Safety at Work Regulations 1992 require all employers sharing a workplace to co-ordinate their activities and to co-operate with each other. The CDM Regulations go a stage further in that it requires the principal contractor to ensure that this happens. The principal contractor should do this both in terms of intercompany involvement and at the workforce level.

At the management level, the principal contractor should ensure that there are formal meetings with each subcontractor which include health and safety on the standard agenda. It is also recommended that a separate forum be established whereby the subcontractors collectively meet the principal contractor, again with health and safety on the agenda.

However, the main effort in securing co-operation should be directed at individuals. The participation, commitment and involvement of everyone on site in health and safety issues is the real goal.

The Safety Representatives and Safety Committees Regulations 1977 provide for the appointment of safety representatives by recognized trade unions and for the formation of safety committees when requested by safety representatives. In recent years there has been a marked decline in activity in this area and it has generally been only the largest of projects which have had formal safety committees.

The Health and Safety (Consultation with Employees) Regulations 1996 top up the 1977 regulations by extending consultation to any employees who are not members of a group covered by trade union safety representatives. These regulations provide for consultation either directly with employees or through elected representatives.

The establishment of a site-wide safety committee and a requirement for all subcontractors to be represented therefore not only enables legal compliance by the various parties, but also signals the principal contractor's commitment to involving everyone. The committee should be chaired by the senior site manager, should be clear in its aims and objectives, and should stay strictly within defined terms of reference. It should not merely be reactive in terms of accident investigation and analysis but should involve itself in the planning and measuring of health and safety improvements.

Other, less formal mechanisms for encouraging participation should also be pursued—for example, the inclusion of question and answer sessions at inductions, briefings, toolbox talks and training sessions, and a clearly stated 'open door' policy on all health and safety issues.

Monitoring

Every subcontractor must supervise his own work in order to manage his own business. This should include his own inspection regime to ensure that he is at least complying with his statutory requirements. The principal contractor's role is to monitor the subcontractor's performance properly and to check what is actually happening against what should be happening.

Monitoring performance, including health and safety performance, is a line management function and how it is to be carried out should be clearly defined by the principal contractor. It is an important activity in itself and also as a clear signal to subcontractors that their health and safety performance is important. It is commonly described as consisting of two elements: active and reactive.

- active monitoring involves checking for compliance with stated objectives and standards—are they doing what they said they would do?
- reactive monitoring involves checking on accidents, incidents, near misses and defect reports.

The precise nature and frequency of subcontractor monitoring by a principal contractor will vary according to the complexity and level of risk of the work, and the competence and degree of self-monitoring by the subcontractor. However, the following items are typical:

- frequent direct observation (several times per day), by the manager with immediate responsibility for the subcontractor, of their work and behaviour, to check compliance with method statements
- weekly examination of the subcontractor's inspection reports and statutory registers
- monthly examination of subcontractor's progress reports, accident statistics, etc.
- checking of subcontractor's training provisions, certificates of competence, etc.

- ☐ checking by direct observation that remedial actions are completed as agreed.

In addition to the above, it is normal practice for health and safety inspection and audit systems to be established by the principal contractor and carried out by their own specialist safety advisers. The emphasis here is gradually switching from the traditional safety inspection of physical conditions to much broader safety audits, the intent being not simply to produce another list of defects but to establish why defects may arise. In order to be comprehensive, audits must automatically encompass not only the work of the subcontractors but also the management of those subcontractors by the principal contractor.

Various types of audit systems have been developed and they may generate both qualitative and quantitative data on health and safety performance. It is recommended that unnecessary complexity or sophistication should be avoided. A consistent approach by competent people is more important. The ultimate aim is to provide an independent assessment of whether the whole of the safety management system is working as intended and actually achieving consistently good results.

Review

One of the most commonly overlooked parts of a comprehensive safety management system is the review process. This is aimed at using all the information from monitoring and audit systems in order to learn from experience and seek continuous improvements.

While review and feedback should be continuous throughout a project, they are particularly important at the end of each subcontractor's package of work. The review should involve evaluating the health and safety performance at all stages of the subcontract process:

- ☐ planning
- ☐ choice of subcontractor
- ☐ execution of works
- ☐ effectiveness of monitoring.

Lessons learned must be recorded and fed back into the overall safety management system such that improvements can be made. Findings relating to the subcontractor's performance should be incorporated into any database maintained in order to assist in the future selection of subcontractors.

The review process should be carried out by those who were responsible for the subcontracted works and may also involve the subcontractor in order to obtain a differing perspective of the process. The results should be copied to the subcontractor in order that he too may learn from the experience.

Summary

The management of subcontractors, or, more correctly, the management of the subcontracting process, is a key issue for the construction industry. The management of health and safety in subcontracted works is an integral part of that issue. The normal business principles of risk transfer do not apply to health and safety issues. Principal contractors are increasingly being held accountable in law for the actions or inaction of their subcontractors.

The subcontracting process has been broken down into planning, choosing subcontractors, working with subcontractors, monitoring and review. At each of these stages overall business performance and health and safety performance are inexorably linked. The same systems and controls are necessary to control risk and therefore to succeed in both:

- identify the problems
- plan the solutions
- communicate these solutions
- implement them
- monitor their application
- review the whole process.

This process of management is as relevant to health and safety and subcontracting issues as it is to any other part of a business.

Organizations wishing to improve the health and safety performance of their subcontractors should look first at their own systems for managing the subcontracting process. Only when these systems are effective will the outcome be assured.

Bibliography

Construction Industry Board (1997). *Code of Practice for the Selection of Subcontractors*. Thomas Telford, London.

10. International safety considerations

Peter E. Brown, *Bechtel Ltd, London*

Introduction

The comfortable framework of legislation and the culture which exists in the UK and across Europe are obvious factors in helping reduce the stress levels of the safety practitioner charged with the responsibility for complying with and applying the necessary safe systems of work. Likewise, supervision and engineers working in the construction industry in Europe can also be considered a key element in the effectiveness of the overall safety management systems being applied, playing their part in the overall compliance to the legislation required in the member states. But what happens when these people move across the boundaries into the international arena, plying their trades in areas where legislation is absent or not enforced? Places where the culture is one where risks that are not acceptable in this country are perceived as normal working practice, and where the working population is made up of every conceivable nationality.

Those people embarking on an overseas venture for the first time will, in all probability, experience some measure of trepidation and nervousness prior to the assignment; one of the concerns being that the safety blanket offered by the systems in place for supporting the safety effort will no longer be there, leaving them both exposed and having to rely on their wits to achieve the best possible standards.

However, this is perhaps an unfair assessment, and the reality is usually more positive. Indeed, it is a fact that, statistically, safety performance on overseas projects can be attained that far surpasses anything experienced in the UK. It is not uncommon for many millions of hours to be worked without lost time injury, and the theories put forward by Heinrich, Bird, etc., on the statistical relationships between fatal and serious accidents, lost time accidents and near misses also bear little or no relationship to the statistical records shown on many projects.

This chapter has therefore been written to give a practical insight into overseas working conditions and the factors that influence the safety performance you can realistically expect on site.

The law!

Health and safety legislation on overseas locations is a variable factor with regard to standards and application. Conversely we should also understand that some of the prescriptive requirements found under the 'six pack' Regulations, and other examples of UK legislation, are completely alien to people coming to work in the UK. Even our American cousins struggle with the idea that they are bound by law to provide basics such as safety footwear when working in the UK—something that is not a duty in the USA. Nor is there a requirement for a risk assessment process and even prohibited substances such as the use of sand for blasting operations are perfectly acceptable. Another classic example is the use of asbestos—perfectly legal and acceptable in India!

So it follows that there is a great divide between the cultures that operate internationally, and the standards of acceptability that are applied with regard to health and safety. This often makes the legal considerations of the host country a secondary consideration, and it becomes very much up to those persons charged with the responsibility for project management how, and to what standards, health and safety rules are applied.

To summarize the legal considerations, it is probably safe to say that outside of those geographical areas such as the Americas, Australia and Europe, legislation and its application is not the driving force to achieve high standards, and we should move on to the areas that really affect safety performance.

Safety standards

It is a well-known fact among safety professionals that good systems of safety save money, enhance schedules and ultimately reduce costs. Unfortunately this view is not universally shared, and as the driving force for construction completion is the project management team, then safety is in a make or break mode with regard to standards of health and safety and the performance expectations. However, this situation is not ubiquitous—as an example of good site management, some of the best scaffolding standards seen by the author were in Malaysia and Thailand erected by locals, who had been trained to Western standards.

Fundamentally there is no excuse for poor safety standards, and yet the following examples of bad practice are evident on many overseas projects:

- site-fabricated ladders made out of scrap material and nailed together
- bamboo scaffolding
- electrical cables strung out as a tripping hazard and with bare wires pushed into a socket—remember also that 110 V tools are not available everywhere
- lack of personal protective equipment—locally made eye protection with glass lenses for example

- ☐ sheer-sided excavations without side supports or where sloping/ benching could have been provided
- ☐ reinforcing bars used as chisels, hammers, crowbars, etc.—it is amazing what 'rebar' can be used for
- ☐ bicycle wheels used a hoist
- ☐ walking the steel
- ☐ poor housekeeping
- ☐ no hygiene arrangements, with personnel eating wherever they can, attracting large numbers of flies
- ☐ no site sanitation, etc.

The list would be endless, but what really lies at the root of these obviously unsafe situations?

Unfortunately when the average expatriate starts working in the Middle East, India, etc., the standards they are used to applying become forgotten. There is a tendency to ignore and avoid the application of good safety standards. This is an observation rather than a criticism, and in many cases the focus for management and supervision is on getting reasonably efficient systems in place with a quality product at the end of it that achieves both schedule and cost targets.

When management and supervision are quizzed as to why safety standards are so poor (any safety professional with overseas experience who reads this may well recognize the statements below) the following clichés are often given:

- ☐ You can't expect the same standards on this project that you get on a job in the UK.
- ☐ These people are not used to the standards you're trying to achieve.
- ☐ We have never had to do that before.
- ☐ If you try and apply the same standards here as in the UK, then the job will stop.
- ☐ We've got to get the job done.
- ☐ Life has a lot less value in this country.

The above comments may sound cynical, but they are familiar excuses often fronted by people who have years of experience, and roughly translated they mean: *I didn't expect the same standards to be applied. These standards meet my expectations and it's going to take all my time if I try to apply high standards of safety.*

This may sound unfair, but it is an unfortunate fact of overseas life, particularly among the smaller contracting companies. Many supervisors experience a change of management style when they have been working the international scene for a few years, and are a major 'negative' influence in dragging standards down.

What's the fix?

The fix is with project management in the planning stage of the project. Aspects such as organization, planning and training are crucial elements to any job, but can take on a different flavour when working overseas.

Organization

The more people you have on a project then higher costs are obviously incurred, but with the add on for working overseas—air fares, meals, accommodation, uplifts, etc.—the costs are driven up considerably. Despite this, a close look at the type of construction project you are engaged on will determine what special areas of expertise are required.

As an example if the work is a petrochemical construction management project, there will be a lot of working at height and lifting operations that need controlling. A scaffold co-ordinator brought on early can train and educate the workforce into the skills necessary for ensuring that site standards are maintained. This person alone can bring a lot of value to the project organization. One project in the Far East had a UK-based scaffolding company brought on to provide all the site scaffolding. The training they gave the local population left them with skills they would never have had otherwise and the standard of scaffolding was exemplary. The project was brought in on time with an impressive record of 35 million hours worked with no fatal injuries, only three lost time accidents recorded and was significantly under budget (more than $50 million). The scaffolding philosophy was a major factor in this achievement.

Whenever there is an accident involving a lifting operation, it is in most cases very serious, very costly and can put a sizeable dent in the schedule. A rigging supervisor takes on an almost quality assurance role along with the training, planning and monitoring activities—setting up the systems for ensuring the competence of operators and riggers, checking the equipment brought to site, working with the engineers in establishing the safety margins for particular lifts. This is not just about safety, it is risk avoidance and goes a long way to promote the success of the project execution.

Planning

Planning issues take on new and varied aspects when applied to an overseas location. For example: how would you plan for a project in Central Africa? In terms of health and safety, working in remote places presents unique and difficult challenges, the following examples are just some of the environmental considerations which should be looked at.

Disease

Places such as Central Africa have a range of diseases and ailments which the majority of people have never heard of. Malaria and other infectious tropical diseases are quite common and together account for 1–2 million deaths a year, but there are also some particularly nasty infections and ailments which, although not so common, are a consideration when planning health and safety on a project.

To give emphasis to this statement let us look at a waterborne hazard that is found in places such as Chad. Guinea worm is spread by water fleas. If ingested, either by drinking or swimming, the flea releases parasite larvae which bore through the intestinal wall into the abdominal cavity. After a period of time the larvae migrate and move into the connective tissue under the skin, where they mate. The male dies and the female keeps growing to a length of up to 30 in., usually locating itself in the legs. When the female is ready to release its eggs into the outside world, it does so by boring a hole through the skin, creating an ulcer. This is an extreme case of a potential health risk, but it clearly demonstrates the fact that newly assigned staff need educating on these types of risks, and this can only be achieved and communicated by previous research and planning.

Local environment

In places which are largely underdeveloped, socioeconomics plays an important role in planning the project. Great care has to be taken that caste systems and tribal differences are understood and planned for. This may sound like an industrial relations issue, and in part it is, but it does have a profound impact on the project safety performance if not handled right.

Considerations should include:

- What is the peak workforce? Are there enough manual workers in the area to resource the project? In tribal areas a huge influx of itinerant workers potentially causing disruption is not a desirable thing to happen. Early consultation can forewarn and prepare for this.
- What is their religion? If Muslim they have the right to pray, and facilities have to be provided to cater for the needs of that section of the workforce.
- What is the opinion of the local population: are they in favour of the project or against it? In some areas the local people can be extremely militant and precautions are required to ensure the safety of the expatriate worker. On one project in India the local community has in the past caused disruption by picketing the gate and throwing rocks at the project employees.

Another instance occurred in Pakistan when members of one sect were employed in the kitchens and gave short rations and scraps to members of another group. This resulted in a riot on the project, with armed personnel coming into the project facilities. Fortunately this was quickly defused and no injuries were sustained, but it does illustrate the point of how important considerations of this nature can be.

Training

Training is one area which has the potential for the greatest impact and can make a major contribution to the project's overall success. The following views expressed are those of the author and, though not proven, appear to be one of the differentiators between UK and overseas performance.

A young person in a developed country who leaves school and takes up a trade is usually taught by the practitioners of that trade. Those practitioners were also taught their trade at some time in their lives, presumably in similar circumstances. So what did they learn? In essence they learned behaviours which are continually reinforced by repetition; these behaviours can be good and they can also put the individual at risk.

When training is applied to these people its effectiveness is diluted, and maybe totally negated because of the inbuilt prejudices that evolved with the experience gathered. Safety induction training often fails because the personnel attending have heard it all before and are not receptive to hearing it again. Any safety practitioner who gives induction training in the construction industry on a regular basis will probably recognize some of the people attending and be quite used to the negative attitudes and comments expressed. This can sometimes be alleviated by written tests but it does not stop or address the real issues of preventing the unsafe act in the workplace. However, there is an interesting aspect to this which is covered by the behavioural section of this chapter, demonstrating the ability of any worker to have a profound knowledge of the hazards in the workplace.

Training on international projects has an entirely different effect. It becomes a far more effective tool for promoting the safe behaviours that you are looking for on a project, helping you work far more hours with far fewer incidents.

Why, when a majority of the working population on a project is totally inexperienced in construction, do these people have significantly fewer accidents than their UK/European counterparts? The answer, in all probability, lies in the fact that the training given is an education in itself—education implying a far more profound understanding of the process than training. These people come to a project without the skills and training found in the European worker. They are in a sense a clean sheet of paper without any of the prejudices, preconceived ideas and habits that lead to accidents, and they are receptive to new ideas and methods of work. Thus the training given is the education process itself, rather than the superficial understanding of site rules and safe systems of work.

Risk reduction

The process of risk reduction is well documented and is an everyday feature of UK law (e.g. Management of Safety and Health Regulation 1992). How do these principles apply internationally when the workers' perception of what constitutes a hazard is totally different to ours. One of the big issues that always rears its head is that of working at height. Thai, Filipino and Indian workers, for example, appear to be totally blasé about the risks associated with working at height and display a confidence that is quite unsettling to Western eyes.

For example, several years ago in Thailand the author saw a high-rise block of flats being built which was about 150 m in height. Hanging from the roof were three rope ladders with single wood boards spanning the ladders and supported by the rungs. At about the 125 m elevation, a man was standing on one of the boards cleaning up the outside of the building. A child of about seven was sitting at his feet with his legs dangling over the edge eating lunch. It should be mentioned that some of the expatriate construction workers had to leave the area because they were so worried by the activities taking place.

However, this is a case in point where the hazards perceived by an individual from one culture were not shared by someone from a different culture. It is quite likely that the risks created by that situation would be totally lost on the persons concerned. This raises an interesting question and something that we should all be mindful of. Accepting that risk is a balance of severity and probability, the probability factor was no doubt considered to be quite low. The thoughts that would run through our minds—the board splitting, a rope failing, slipping and tripping, etc.—would not be a consideration in this case, enforcing the need to educate and create the greater awareness we all should have when planning the work.

The above example is typical of the many that occur on overseas construction sites, and though the at-risk behaviours are more frequent in terms of exposure over a period of time, the number of accidents that occur are relatively few. It has to be said, however, that when they do happen they are predictably quite severe.

The 300:29:1 ratio put forward for the correlation between minor, lost time and fatal/serious accidents (based on UK/US performance analysis) tends to go haywire on the international project. In extreme cases you can have a handful of first aid cases, one or two lost time accidents and then something as serious as a fatal, so the correlations and trends cannot be predicted with the same certainty.

Accident prevention

Having said that, the correlation and ratio between accidents is not fixed, there is something that can be done, namely identification of the common causes of accidents. We have already looked at issues such as training, organization and planning and they are, in truth, the principle reasons behind many of the accidents that do get reported.

A recent incident occurred when the staircase and standing platform on an overhead crane failed and collapsed, killing the man who had been standing on it. The accident really happened 17 years previously when the design specifications were not observed and the quality control procedures failed. In fact the platform should have extended under the crane operator's cab with continuous welds applied at the two points where it interfaced with the cab body. It ended up as being welded to the side of the cab by several tack welds and was a component failure waiting to happen.

This example demonstrates that accidents have many possible causes and that on international projects there may well be a tendency not to follow procedures, to take shortcuts, etc. It emphasizes that what happens in the engineering and project management offices can have an effect and make the difference.

Selling safety

We have covered some of the major factors to be considered with regard to overseas work, but there is another issue that should be thought about, namely 'How should a safety message be sold'? An amusing feature of project life is the diverse range of safety signs that proliferate on a project. The Koreans are particularly good at having cartoon safety figures on sign boards—usually bearing the legend 'Safety First', which in reality may translate into 'Safety Last' when examining site practices!

Selling the safety message in the right way is extremely important and you have to have a clear understanding as to the cultures you are dealing with. A 'sword of honour' is often used as an award for safety, but the presentation of a sword would be considered an insult to the Chinese.

Knowing your audience sounds a simple exercise, but it is important to consider literacy levels and what people like/dislike. In Thailand, for example, the message is best sold by using graphics with humorous overtones. The Thai people have a keen sense of humour and like high-visibility material; so giving them what they can identify with gives more power to what you are selling. The use of graphics is also a preferred mechanism where multiple ethnic groups are working. Singapore is a prime example: while investigating one incident it was found that we had a Thai welder supervised by a Chinese foreman taking directions from a Malay supervisor—how did those guys communicate?

It would be impossible to cover all the scenarios you may encounter, and basically the advice is to get the understanding required to be effective in the working environment you may find yourself in.

As a final point, I would ask that you consider the messages that are displayed on project target boards and safety posters. In most instances they will have a negative message—number of days since the last lost time accident/how many accidents sustained that year and where they happened and to what contractor. Sometimes companies will use body parts, highlighting the number of incidents that happened to the head, hands, feet, etc. Posters showing people falling off ladders, being transported to hospital covered in blood are also typical. These are very negative and do nothing to sell the positive message. Saying you have had an accident or even depicting an accident is a sign of failure and not setting a challenge that the workers can aspire to.

Behaviour-based safety programmes

If you quiz a safety professional on the examinations he has to take to be qualified, he will undoubtedly tell you about safety management, law, occupational health, general science and behavioural science. Behaviour-based safety programmes are starting to get recognition and the thrust in the UK has been to move away from process conditions, to start homing in on the people and their behaviours in the workplace.

So, are behaviour-based programmes too sophisticated for international projects? Basically the answer is 'no', because there is nothing particularly difficult with this concept. The underlying philosophy is to give ownership of safety to the people at risk. Traditionally projects are managed by people who are not at risk giving direction to those who are. It is a major change to acknowledge the role of the individual in the workplace and to invest in that person a sense of control and ownership of his actions. Supervisors also have to accept that they now play a more proactive and positive role. Rather than being in total control, they now play a lead figure in a team context.

The way in which these programmes work is to monitor workplace behaviour and to identify the 'at-risk' behaviours and the 'safe' behaviours. The persons who do the monitoring are not safety professionals or members of supervision, but people who are selected from the workforce or from non-threatening disciplines such as nursing staff. They create the checklists that measure behaviour (working from ladders, compliance with Personal Protective Equipment, improvising tools, etc.). The results of this 15/20 minute survey are discussed with the employee, with both positive and negative aspects being addressed. The results of all these surveys are fed back to the project organization in a manner that measures the 'at-risk' trends. The whole cycle is completed when an action plan is developed that targets the number of 'at-risk' behaviours on site and the effort required to reduce those trends.

Sounds simple: it is! This simple tool is underpinned by the basic philosophy of giving people their place, and respecting them for the impact that they can have on your safety programmes. It breaks down many of the traditional barriers found on overseas projects and fosters the team environment—something which is seldom seen internationally.

Caution must be exercised, in that cultural considerations such as caste, language, hierarchy, etc., are considered, and the approach tempered accordingly. It can be done and there are many benefits from having a well motivated and committed workforce who actively participate in the safety programme and are not just bystanders.

Conclusion

Irrespective of where you are in the world, nobody goes to work to be hurt, and companies are not in business to harm their employees. The safety and health of the worker is an important issue and the programmes designed for their protection should never be considered as an 'add-on' to the construction execution programme, but closely integrated into every aspect of the work to be managed.

Hopefully after reading this chapter, you should have an understanding that safety performance is an attainable goal on international projects; that it does bring value to the project; and that it should also be considered a *personal value* to the people managing these projects—when that happens the drive towards a zero-accident culture becomes a meaningful and realistic target.

I hope the messages and information conveyed in this chapter are of use and that, for the people embarking on an overseas project, their perception is one of challenge and opportunity, rather than trepidation. I have worked with many nationalities and made many friends in the process from Korea, Thailand, Japan, New Zealand and the USA. The list is long, but how fortunate I have been to have had that opportunity, and to learn about different cultures and to work with these people in the area of safety. I have had support from the most unexpected quarters and visibly seen the results in improved workplace conditions and reduced accident frequency rates.

11. The Health and Safety Commission and Executive: dealing with an inspector

Stuart Nattrass, *formerly Health and Safety Executive*

Origins of the Health and Safety Commission and Executive

The Health and Safety Commission (HSC) and the Health and Safety Executive (HSE) were created by Section 10 of the Health and Safety at Work etc. Act 1974 (HSWA).

The HSC/E regulate almost all aspects of industrial safety in the UK. However, they do not deal with consumer and food safety, marine and aviation safety and pollution. This wide scope means that several government departments are responsible for considering proposals for new law and standards. However, the main ministerial link is with the Secretary of State for the Environment and that Department.

Place of the HSC/E in the health and safety system

HSC/E are the prime movers in the UK health and safety system. The principles on which the system is based help to explain the make-up of HSC/E and the activities which they undertake.

The first principle is that primary responsibility for identifying hazards, assessing risks, implementing controls and monitoring whether they are adequate lies with industry. In order to secure this degree of commitment, it is essential to involve employers, employees and the wider community in making policy and setting standards.

Another principle follows from this. Decisions about tolerable levels of risk are not matters for experts alone. Experts have an important contribution to make and it is their job to give an informed opinion as to levels of risk and the extent to which they could be reduced and at what cost. The HSC/E do not have a monopoly of expertise and it is important to draw on all available expertise, whether it is in professional bodies, research or academic circles, or trade unions, etc. However, decisions about tolerability of risk ultimately are political matters, not expert ones. When experts have made their recommendations, society has to decide what it regards as tolerable risk and cost. These decisions are made by Parliament in the final analysis but in practice through the HSC.

Another important principle is that people need information about health and safety in order to play their respective part in the system. This applies to employers, employees and members of the public who may be affected by work activities.

The final principle is that some external co-ordination, stimulation and control is needed. A body has to hold the ring in order to ensure that important hazards are identified, standards set, and information and guidance produced. A body is also needed to check that agreed standards are being achieved in workplaces. These functions are performed by HSC/E, together with local authorities in their enforcement capacity.

Make-up of the HSC

The HSC has a Chairperson and nine members, appointed by the Secretary of State for the Environment. Three members are appointed after consultation with organizations representing employers. Organizations which represent employees are consulted about another three members. Other members are appointed after consultation with local authorities and other bodies which represent public and professional interests, etc.

Functions of the HSC

The HSC's functions and powers are set out in Sections 11–14 and 16 of the HSWA. In summary they are to make arrangements to secure the health, safety and welfare of people at work and the public from the way undertakings are conducted. This includes proposing new law and standards, conducting research, and controlling explosives and other dangerous substances. It also has a general duty to help and encourage people concerned with these matters. Together with the HSE, the HSC publishes annual plans of work and reports.

The HSC consults widely before proposing new law and standards. It issues consultative documents and also maintains several advisory committees which deal with particular hazards or industries. In the case of construction, the Construction Industry Advisory Committee (CONIAC) advises the HSC. All these committees include a balance of employers, employees and technical and professional experts.

Make-up of the HSE

The HSE is legally a body of three persons appointed by the Secretary of State. The Executive is the main instrument of the HSC, which it advises and supports. Under Section 18 of the HSWA it has a specific responsibility to make adequate arrangements for enforcement except where other arrangements are made under regulations. The Health and Safety

(Enforcing Authority) Regulations 1989 make local authorities responsible for enforcement in various sectors, generally retail, commercial and leisure, except where local authorities themselves are also the employer or are in control. In the case of construction, some minor work is inspected by local authorities but the great majority is inspected by the HSE. The Commission may not give to the Executive any directions as to enforcement in any particular case.

The Executive is the employer of inspectors, policy advisers and technical, scientific and medical experts—around 4000 people in all. The legal body is known as 'the Executive' and the organization as the 'HSE'.

Organization of the HSE staff

The HSE staff fall into four general groups:

- ☐ Policy staff advise the Commission on new law and standards, liaise with government departments and consult on the Commission's behalf.
- ☐ Inspectors enforce health and safety law. Manufacturing industry, services, construction, agriculture and quarries are covered in regions by the Field Operations Directorate (FOD). Railways, nuclear installations, offshore installations and chemicals, and onshore hazardous installations are covered by separate divisions, some operating regionally and others centrally. Most construction work is inspected by the FOD but, as a rule of thumb, other divisions inspect where the construction work is not set apart from the main activity.
- ☐ Technical, scientific and medical staff advise on best practice and are responsible for technical standards and research. The Employment Medical Advisory Service (EMAS) is responsible for advising about health in relation to employment. It operates within the FOD regional structure, together with specialist staff who provide technical support to inspectors.
- ☐ Information technology and other staff provide common support services to the whole organization.

Structure of the HSC/HSE

The structure of the HSC and the HSE are shown in the chart in Appendix 1.

HSE activities

To achieve its purposes, the HSE undertakes a range of activities. Some activities are directed at industry generally, or at particular sectors. Other activities are directed towards particular employers.

Activities directed generally

Generally directed activities include:

- The publication of Approved Codes of Practice and guidance, both priced and free. They can be ordered from HSE Books, PO Box 1999, Sudbury, Suffolk CO10 6FS, tel. 01787 881165 or fax 01787 313995.
- Participation in setting standards, prepared by standards-making bodies such as the British Standards Institution (BSI) and European bodies such as CEN and CENELEC. Frequent reference is made to appropriate standards in HSE published guidance. The HSC/E are themselves standards-making and standards-certifying bodies.
- A national telephone public enquiry service, HSE Infoline. This answers general queries and advises about publications. It can be contacted by telephone on 0541 545500.
- The HSE also has information centres for personal callers in London (Rose Court, 2 Southwark Bridge, London SE1 9HS), Bootle (St Hugh's House, Stanley Precinct, Bootle L20 3QY) and Sheffield (Broad Lane, Sheffield S3 7HQ). Written enquiries may also be addressed or faxed (0114 289 2333) to the Sheffield centre.
- An HSE Gas Advice Line, which gives advice about domestic gas safety, on freephone 0800 300 363.
- Working through intermediary bodies to increase the awareness of duty holders, for example through TECs and LECs, chambers of trade and commerce, trade federations and trade unions.

Activities directed towards particular duty holders: inspection

The inspection activity is directed towards particular duty holders, to see whether and how they are discharging their duties and making use of all the information and advice which is generally available.

The primary objective of inspection is to stimulate employers and other duty holders to carry out their duties to provide and maintain safe and healthy working arrangements.

Powers of inspectors

In carrying out inspection, inspectors use powers which are given under Section 20 of the HSWA and they have a warrant to this effect. These powers are summarized in the HSE's free booklet *Working with Employers*, which explains that inspectors have powers to:

- enter premises at any reasonable time (or at any time if they think a situation may be dangerous)
- carry out examinations and investigations; take measurements, photographs and samples

- ☐ take possession of an article and arrange for it to be dismantled
- ☐ seize and make safe any article or substance believed to be a cause of imminent danger or personal injury
- ☐ require information and take statements from people they think can help an investigation
- ☐ inspect and copy documents
- ☐ issue improvement and prohibition notices and prosecute.

Types of inspection

There are two main types of inspection. The majority are initiated by the HSE as part of a programme of planned preventive inspections. Reactive inspections are in response to accidents, dangerous occurrences, diseases and complaints.

Purpose of programme of preventive inspections

The HSE's overall purpose in carrying out a programme of preventive inspections is:

- ☐ to identify and monitor the most hazardous activities, for example nuclear and other hazardous installations or work, irrespective of the competence of those in control
- ☐ elsewhere to identify poor performers and follow them up until they comply to an acceptable level
- ☐ to scan the rest of the field sufficiently frequently to keep duty holders motivated by generating a reasonable chance of detection and also by assisting and encouraging the reasonable performers.

In order to meet these aims, the HSE has a prioritized programme of visits which ensures that inspectors go most frequently to places most likely to need inspection. The factors on which priorities are based include the level of risk to employees and the public (with appropriate allowance for low-probability/high-consequence events such as might arise from major hazard activities), the commitment of management and their performance as shown at previous inspections, and the length of time since the last inspection.

Purpose of particular inspections

The purpose in visiting a particular duty holder is to determine whether there are systems to ensure effective management of health and safety, covering policy, organization, planning and setting standards, implementing and monitoring the use of controls, and review. Therefore the inspector will wish to have the systems explained, with documentary evidence. Selected parts of the work activity will also be inspected to check that the system is working in practice and adequate controls are actually in place.

In the case of construction, these principles apply equally whether the inspector is visiting a site during the construction phase or a design office during the preconstruction phase.

Thus at the start of an inspection, the best analogy for the role of the inspector is that of an auditor. The inspector will approach the inspection on the assumption, until the contrary is proved, that he or she is dealing with reasonable people who are attempting to discharge their legal duties.

Methods of inspection

HSE inspectors generally visit without giving notice, although an appointment may be made when it is essential to meet a particular person or see specific documents. The method of inspecting will vary, depending on the size and complexity of the work organization. In some cases the inspection may be carried out by a team of inspectors, but in most cases the inspector will be unaccompanied. However, in all cases the same basic principles apply. These are that the inspector or team will ask to meet the person in charge and responsible for health and safety. They will ask them to explain the company's approach to health and safety and to produce the health and safety policy and other relevant documents. For example, in construction the preconstruction or construction phase plan may be requested if the Construction (Design and Management) (CDM) Regulations 1994 apply. Other documents that may be requested include the record of significant findings of risk assessments under the Management of Health and Safety at Work Regulations 1992. These will often be associated with method statements.

The person in charge will be invited to say what they regard as the most significant risks and what steps are being taken to control them (or avoid or reduce them in the case of the design stage). Selected parts of the activity will be inspected subsequently, but no inspection tries or claims to see every aspect of the work. It is essentially an audit based on a reasonable sample, covering the most sensitive aspects. However, the principle remains throughout that responsibility for ensuring safety rests with the duty holder and not with the HSE or the inspector.

At the end of the inspection, the inspector will say whether improvements are needed. For routine matters they will give general advice on how improvements may be made, but they will not act as an in-depth consultant. Their advice may be confirmed in writing. Where serious failures are identified, the inspector may use formal powers of enforcement (improvement and prohibition notices and prosecution—see below).

The HSE operates on the principle that duty holders are entitled to an explanation of an inspector's judgement and proposed line of action and may ask to see an inspector's manager if they disagree.

Consultation with employees' representatives

An important aspect of an inspection is the inspector's duty to involve and inform representatives of employees. During the visit the inspector will ask to meet them privately to obtain their impressions about the adequacy of health and safety management and to enquire whether there are any serious matters which they consider should be covered during the inspection. At the end of the inspection, the inspector will inform the representatives what has been discovered and what action is proposed. The same information will be given to the employer. The inspector may send a confirmatory letter to representatives.

Reactive inspections

The same principles apply to reactive inspections, that is the investigation of accidents, dangerous occurrences, diseases and complaints. While the inspector will wish to determine the immediate cause of the accident, the main focus of interest will be the underlying cause. Therefore the inspector will take the accident as a sample of the overall way in which health and safety are managed. Employees' representatives will be involved in the same way as in preventive inspections.

Investigation of accidents, dangerous occurrences and diseases

The HSE does not attempt to investigate all reported accidents, etc. Fatal accidents are always investigated. Otherwise, selection is according to whether there appears to have been a serious legal contravention or whether any new information is likely to emerge as to causation and prevention.

Investigation of complaints

The HSE has a policy of investigating a high proportion of complaints. When a complainant contacts an HSE office, they are asked whether they wish their complaint to be treated as confidential. Unless a complainant gives permission, the fact that a complaint has been received and its source are not disclosed. The way in which a complaint will be investigated is discussed and agreed with the complainant, for example by a special visit or at the next inspection. The result of the investigation is notified to the complainant. Before the HSE agrees to investigate a complaint, the complainant is asked whether they have taken up the matter with employees' representatives. The HSE will not usually intervene until internal mechanisms for resolving concerns have been exhausted.

Confidentiality and disclosure of information

The HSC has published its policy on access to information in a free booklet entitled *Policy Statement on Open Government*. The HSE booklet *Working with Employers* summarizes the policy. It explains that the HSE is bound by various limits on the disclosure of information obtained from businesses. How much they can pass to others is controlled by the HSWA, the Environmental Information Regulations and the Code of Practice on Access to Government Information. In general the HSE will pass on information only when this would help to protect the health or safety of people at work or the public. They will always make information available where it is needed for the immediate protection of someone's health or safety. When such information is passed on, the HSE aims to protect commercially confidential information and also personal privacy, medical confidentiality and national and public security. Where the HSE receives a request for information which may be commercially confidential, they ask the duty holder for advice before deciding how to respond.

Enforcement policy

As explained, enforcement, or securing compliance, is achieved through both informal means (giving information and advice) and formal ones such as improvement and prohibition notices and prosecution. The HSC has published a free leaflet entitled *Enforcement Policy Statement*. It applies to both informal and formal enforcement and contains several principles.

The first principle of the policy is proportionality: this means relating enforcement action to the risks and to the seriousness of any breach of law. The second principle is consistency of approach: this means taking a similar approach in similar circumstances to achieve similar ends. The third principle is transparency: this means helping duty holders to understand what is expected of them and what they should expect from the HSE and other enforcing authorities such as local authorities. The fourth principle is targeting: this means making sure that inspection is targeted primarily at those whose activities give rise to the most serious risks or where the hazards are least well controlled.

Improvement notices

An improvement notice requires that a fault should be rectified within a set time limit. The inspector will normally discuss the time limit with the duty holder. On receipt of a notice, a duty holder may appeal to an Industrial Tribunal. The notice is suspended until the hearing of the appeal.

Prohibition notices

A prohibition notice may be immediate or deferred. The recipient may appeal to an Industrial Tribunal but the notice remains in force pending the hearing of the appeal unless the Tribunal suspends it.

Both types of notice will contain, or have attached, an explanation of what must be done to comply. However, recipients are allowed to use an equally effective alternative method.

Prosecution

In England and Wales, the HSE decides whether to prosecute and an inspector may conduct the case personally in a magistrates' court. In Scotland the Procurator Fiscal decides. The enforcement policy expects that prosecution will be considered when:

- it is appropriate in the circumstances to draw general attention to the need for compliance with the law and the maintenance of standards required by the law, especially where a prosecution would be expected normally or where others may be deterred from similar failures to comply through the conviction of offenders
- or there is judged to have been potential for serious harm arising from the breach
- or the gravity of the offence, taken together with the general record and approach of the offender warrants it, for example apparent reckless disregard for standards, repeated breaches, persistent poor standards.

The same principles apply to the prosecution of companies and individuals, including company directors, managers and employees.

Manslaughter

Where a breach leads to a work-related death, enforcing authorities have to consider whether the circumstances might justify a charge of manslaughter (culpable homicide in Scotland). HSE inspectors liaise with the police, coroners and the Crown Prosecution Service (CPS). If they find evidence suggesting manslaughter they pass it on. If the police or CPS decide not to pursue a manslaughter case, the HSE pursues a health and safety prosecution if it is appropriate. In Scotland, responsibility for investigating sudden or suspicious deaths rests with the Procurator Fiscal.

Making contact with the HSE

General enquiries should be directed to the HSE Infoline (tel. 0541 545500).

Enquiries about specific projects or sites should be directed to the HSE office covering the place where the site will be, or is, situated. A list of addresses is given in Appendix 2.

Notifications of new projects which are required by the CDM Regulations should also be sent there, together with notifications of reportable accidents, dangerous occurrences and diseases.

In each region the HSE's Field Operations Directorate has teams of inspectors who deal with construction. Each team consists of several groups, with a principal inspector in charge of each group.

Points to remember about a visit by an HSE inspector

- ☐ The inspector should be able to identify him- or herself, if necessary by showing a warrant.
- ☐ The inspector will wish to have an explanation of the systems for managing health and safety.
- ☐ The inspector will need documentary evidence.
- ☐ The inspector will wish to inspect selected parts of the work activity.
- ☐ The inspector will wish to talk to key managers and supervisors.
- ☐ The inspector will wish to meet employees' representatives.
- ☐ At the end of the inspection the inspector will say what improvements are needed and whether these are to be confirmed in writing, and also what other action he or she intends to take.
- ☐ The inspector will also give employees' representatives the same information.
- ☐ Whether or not the inspector intends to confirm in writing, it is good practice for the person in charge to make a note of actions which are needed.
- ☐ The inspector will discuss priorities and timescales for improvements, with reasons for them.
- ☐ If the person in charge disagrees or wishes to have a further explanation, he or she may approach the inspector's manager.

If an immediate prohibition notice is issued, it takes effect on receipt and remains in force until any appeal has been heard, unless a Tribunal suspends it pending an appeal.

Appendix 1: structure of the HSE

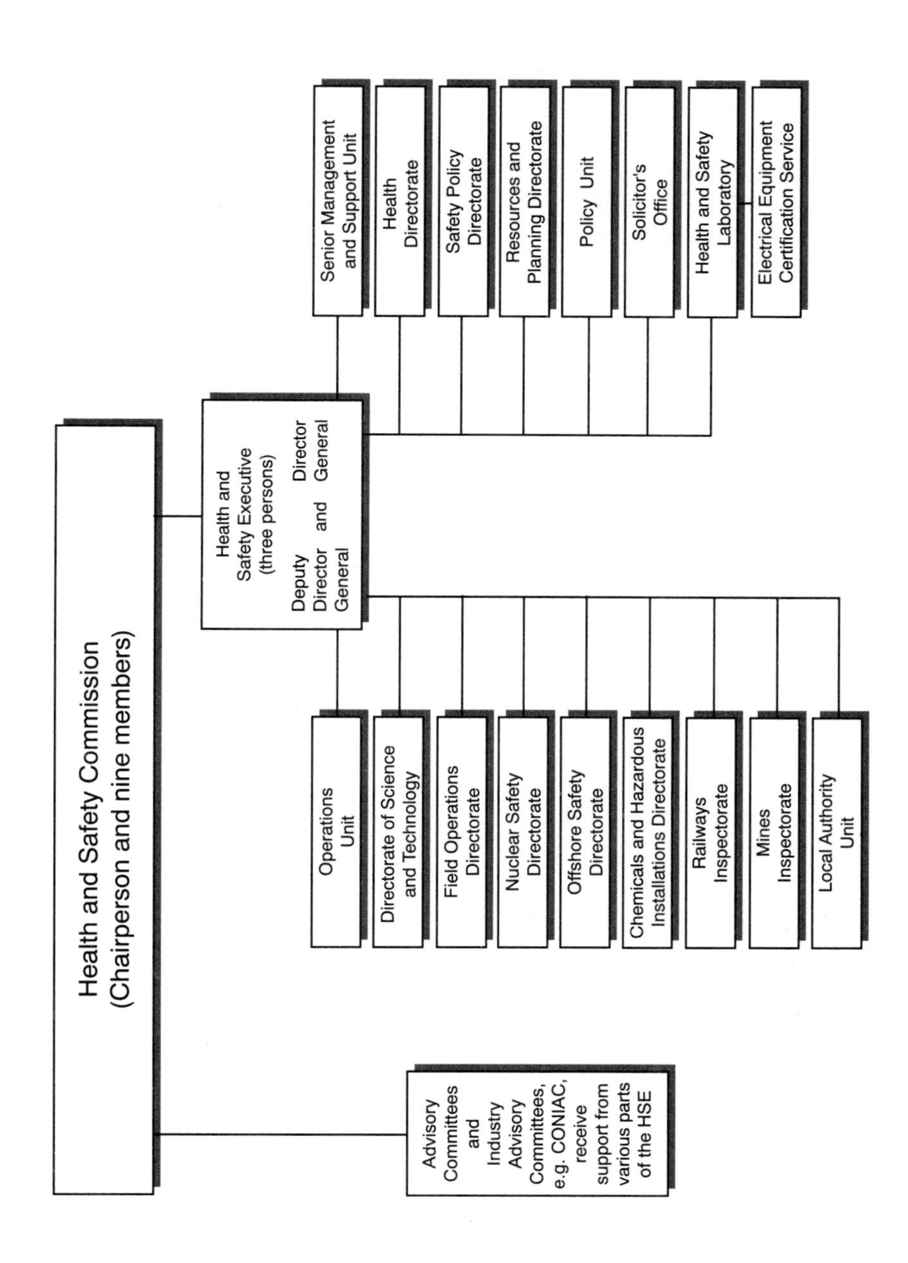

Appendix 2: HSE office addresses and telephone numbers (from Health and Safety Executive, 1996/97)

Wales and West Region

Covers Wales and Cornwall, Devon, Somerset, North West Somerset, Bath and North East Somerset, Bristol, South Gloucestershire, Gloucestershire, Hereford & Worcester, Shropshire and Staffordshire.

Regional office:
Brunel House
2 Fitzalan Road,
Cardiff CF2 1SH
Tel.: 01222 263000
Fax: 01222 263120

Other offices:
Bristol
Tel.: 01179 88600
Newcastle under Lyme
Tel.: 01782 602300

Home Counties Region

Covers Bedfordshire, Berkshire, Buckinghamshire, Cambridgeshire, Dorset, Essex, Hampshire, Isle of Wight, Norfolk, Suffolk and Wiltshire.

Regional office:
14 Cardiff Road,
Luton LU1 1PP
Tel.: 01582 444200
Fax: 01582 444320

Other offices:
Basingstoke
Tel.: 01256 404000
Chelmsford
Tel.: 01245 706200

London and South East Region

Covers Kent, Surrey, East Sussex, West Sussex and all London boroughs.

Regional office:
St Dunstan's House,
201–211 Borough High Street,
London SE1 1JA
Tel.: 0171 556 2100
Fax: 0171 556 2200

Other office:
East Grinstead
Tel.: 01342 334200

Midlands Region

Covers West Midlands, Leicestershire, Northamptonshire, Oxfordshire, Warwickshire, Derbyshire, Lincolnshire and Nottinghamshire.

Regional office:
McLaren Building,
35 Dale End,
Birmingham B4 7NP
Tel.: 0121 607 6200
Fax: 0121 607 6349

Other offices:
Northampton
Tel.: 011604 738300
Nottingham
Tel.: 01159 712800

Yorkshire and North East Region

Covers Hartlepool, Middlesbrough, Redcar and Cleveland, Stockton-on-Tees, Durham, Hull, North Lincolnshire, North East Lincolnshire, East Riding, York, North Yorkshire, Northumberland, West Yorkshire, Tyne & Wear, Barnsley, Doncaster, Rotherham and Sheffield.

Regional office:
8 St Paul's Street,
Leeds LS1 2LE
Tel.: 0113 283 4200
Fax: 0113 283 4296

Other offices:
Sheffield
Tel.: 0114 291 2300
Newcastle upon Tyne
Tel.: 0191 202 6200

North West Region

Covers Cheshire, Cumbria, Greater Manchester, Lancashire and Merseyside.

Regional office:
Quay House,
Quay Street,
Manchester M3 3JB
Tel.: 0161 952 8200
Fax: 0161 952 8222

Other offices:
Preston
Tel.: 01772 836200
Bootle
Tel.: 0151 479 2200

Scotland

Main office:
Belford House,
59 Belford Road,
Edinburgh EH4 3UE
Tel.: 0131 247 2000
Fax: 0131 247 2121

Other office:
Glasgow
Tel.: 0141 2275 3000

Bibliography

Health and Safety Commission (1994). *Policy Statement on Open Government.* HSE, Sudbury, IND(G)179(L).

Health and Safety Commission (1997). *Enforcement Policy Statement.* HSE, Sudbury, MISC 030.

Health and Safety Executive (1996/97). *Working with Employers.* HSE, Sudbury, HSE 35.

12. Twenty-one years on: construction and trade union safety representatives

Tom J. Mellish, *Trades Union Congress, London*

Introduction

In the construction industry 6440 employees and 8300 self-employed out of every 100 000 suffer a workplace injury (Labour Force Survey, 1993/94–1995/96). This is 40% higher than the national average.

Four million working people every year suffer injury or ill-health as a result of their job. The cost of that record to the British economy has been estimated at up to £16 billion (Davies and Teasdale, 1994). But for our members, the record means pain, lost wages and perhaps even the abandonment of their career.

While it is not possible to estimate the proportion of these costs to construction, it must be remembered that construction rates second only to mining in terms of risk. In terms of reportable accidents it lies fourth if using the Reporting of Injuries, Diseases & Dangerous Accidents Regulations (RIDDOR) but unfortunately comes first when using the Labour Force Survey.

This also explains why our members put health and safety at the top of the demands they put on trade unions. Recent research has shown that over 70% of new trade union members consider health and safety to be a 'very important' issue for trade unions to bargain over (more than cited improved pay) (*New Statesman and Society*, 18 November 1994). So we are committed to developing better safety standards at work. Better safety will mean fewer injuries, less ill-health *and* a more productive economy.

The TUC continues to play a major part in the work of the Health and Safety Commission and the British and international bodies responsible for the regulation of health and safety at work. But regulation is not the only way to improve safety standards. Action at workplace level is vital. The TUC has made a firm commitment to working in partnership with employers to improve safety at work. That partnership demands a stronger role for workplace safety representatives.

Safety representatives are backed by legal rights under the Safety Representatives and Safety Committees Regulations 1977, the Management of Health and Safety at Work Regulations 1992, as well as the 1989 European Framework Directive for safety. They can play a key role in ensuring the implementation of workers' rights to a healthy and safe working environment, and in ensuring that workers are fully involved in the development of a safety culture in their organization.

Twenty-one years after the 1977 Regulations, safety representatives have attained their 'majority', and the TUC believes that now is the time to extend their role in the safety system.

Our opinion polling suggests that the British people agree. When asked: 'Do you think people at work should or should not have the right to be represented by a trade union if they want on health and safety?' a staggering 98% of respondents said yes! (TUC opinion poll conducted by NOP, based on 1002 telephone interviews between 6 and 8 January 1995).

This chapter examines the future of safety representatives, setting out how the context in which they work has developed. The TUC wants to see a public debate about how safety representatives can contribute to better safety standards at work. There is an urgent need for that debate in the construction industry. It is hoped that this paper can provide the background for that debate in the construction industry.

Union support for safety representatives

The Health and Safety at Work etc. Act 1974 was a landmark in British health and safety law for three main reasons. First, it created a framework of legislation that set goals, rather than elaborating systems to deal with specific safety hazards. Secondly, to assist in achieving the goals set out in regulations, it established a Health and Safety Executive (HSE) to draw together the work of various inspectorates, and to involve the social partners. And thirdly, it created a network of workplace safety representatives to ensure that workers were actively involved in safety.

The Act originally envisaged safety representatives for all workplaces above a certain size, whether there was a union present or not. The Labour Government, however, amended the Act to ensure that only where unions were recognized would there be legally established safety representatives, and the Safety Representatives and Safety Committees Regulations 1977 confirmed this position.

The reason for confining worker safety representation to union-recognized workplaces was in part a pragmatic one. Without union recognition, it was argued, safety representatives would be unable to ensure that their rights were enforced and their role respected. Although there is nothing to stop employers arranging safety representation for their workforces in the absence of a legal right under the 1977 Regulations, it is more common now for employers, while derecognizing unions for collective bargaining, to retain recognition for safety. Without union recognition, safety representatives would lose access to the support that union membership offers, and they would have less power to ensure that safety decisions were impressed upon management.

Unions offer safety representatives a number of tangible supports. The first is training: since 1977, the TUC has trained over 125 000 safety representatives on the basis of two 10-day courses, with specialist courses of 1 or 2 days duration on specific issues, often organized on a local, sectoral basis.

Individual unions also offer such courses, although not normally of the same duration. Estimates from the Department of Employment's Labour Force Survey suggest that unions provide almost as many places as the TUC—on average, 7000 safety representatives a year are trained by the TUC and the same number by individual unions. Union courses are obviously more likely to have a sectoral focus, and many are delivered jointly with employers, often using TUC materials.

The TUC makes use of 84 centres for quality safety representative training, which is the most comprehensive network of safety trainers in Britain. Colleges of further education and Workers' Educational Association tutors are validated by the Department for Education and Employment (DFEE). Some unions have their own training establishments in addition to these.

Union safety representatives are able to call on a substantial body of information about workplace hazards and the safety systems needed to deal with them. There are at least 250 000 safety representatives in Britain, and many unions maintain regular communication with them. Such union contacts provide safety representatives with access to a trade union perspective on safety matters, through either special mailings, items on safety in union journals or regular safety representative information services.

Union conferences and other gatherings often concentrate on safety matters: there will almost always be a debate about a safety issue at the annual conference, and branch meetings usually feature some focus on health and safety hazards. Safety representatives also obtain information through less formal channels, such as labour research and the hazards movement provided by local safety activists.

A number of unions have gone so far as to establish national structures to deal with health and safety at work. These may be similar in nature to the consultative structure operated by the Engineers and Managers Association whereby their local safety representatives meet on a regional basis and some of them may be elected to their national Safety, Health and Welfare Committee. The General Municipal Boilermakers (GMB) has designated one full-time officer in each region with special responsibility for health and safety at work (supporting the work of other officers), while UNISON, the public sector union, has a lay health and safety officer in each branch who acts as senior safety representative. The TUC's network of union health and safety specialists brings together the national safety officers of all the unions with more than 15 000 members (about 50 unions are involved) in bimonthly meetings. Union representatives on the Health and Safety

Commission's Construction Industry Advisory Committee (CONIAC) play an active role in the union safety specialist network.

Finally, there are a number of national and regional joint bodies outside of the Health and Safety Commission (HSC) structure linking union and employer representatives. A number of these operate at the company level as well.

Union safety representatives and industrial relations

Such joint union–employer bodies also demonstrate the second major strength of union safety representatives—their ability to make use of industrial relations structures and behaviour. Since safety representatives were established, there has been a more or less accepted view that one of their goals should be to take safety decisions out of the realm of normal industrial relations. This has benefits for both the workforce and employers. It should ensure that decisions about safety are not taken for non-safety reasons, and that, therefore, advances in safety cannot be bargained away against wage increases or bonus payments (for example) or vice versa.

But in another sense, the fact that safety representatives are appointed by unions, and only in recognized workplaces, at least until recently, makes them a part of the industrial relations scene in any industry. The opportunity to appoint safety representatives and the legal rights and organizational resource that flow from their status can be a powerful incentive for workers to secure union recognition. And at the same time, their position within their union provides safety representatives with the support of their union's industrial relations structures if they come into conflict with management.

Unions have always argued that without the support of unions, safety representatives would, in the prevailing British climate of industrial relations, be unable to ensure their independence of management. As a result, some unions insist that the safety representatives they appoint should be worker representatives more generally (shop stewards, for instance).

Sources of support for non-recognized workplaces

There are other sources of support for workers who need to deal with a safety problem at their workplace. Most importantly, the HSE will act to deal with any workplace whether unionized or not, and inspectors will (as appropriate) deal with concerns reported anonymously. The HSE's publications are generally available, although increasingly they are priced at a level which makes them affordable more to management than individual workers. There is also a general information point which receives large numbers of calls from the general public, including union members and safety representatives.

Help is also at hand from bodies such as Citizens Advice Bureaux (although they would probably accept that they are less likely to be able to assist enquirers than unions or the HSE) and a small network of Hazards Centres and similar bodies, such as the Construction Safety Campaign, often strongly connected with trade unions. This leaves the majority of workplaces, however, with little access to support over health and safety issues, and the scale of this problem has changed substantially since 1977 when safety representatives were introduced.

Union penetration in the construction industry has declined substantially and now stands at approximately 20%. There are about 3500 trade union safety representatives, the majority on the larger and therefore usually organized sites.

Safety representatives in law

The legal rights of safety representatives were set down primarily in the 1977 Regulations, as amended to take account of the European Framework Directive. Basically, their rights are as follows:

- ☐ to investigate potential hazards and dangerous occurrences at the workplace, and to examine the causes of accidents
- ☐ to investigate complaints by any employee they represent relating to that employee's health, safety and welfare at work
- ☐ to make representations to the employer on general matters affecting health, safety and welfare at work
- ☐ to carry out inspections, to represent workers to HSE inspectors (or other enforcing authorities) and to receive information from such inspectors
- ☐ to attend joint employer–union Safety Committees (which have to be established where requested by safety representatives)
- ☐ to receive adequate training to fulfil their tasks
- ☐ to have access to facilities and time off without loss of pay to carry out the above functions
- ☐ to receive information and take part in consultation, in good time, on a wide range of health and safety issues.

Until the 1996 employee consultation regulations, safety representatives could only be appointed by recognized trade unions, and should normally be employees of the organization whose workers they represent (this is not true in the theatre profession, though). Some have argued that unions should be able to appoint anyone to act as a workforce's safety representative, and this particularly applies to small workplaces, where unions have occasionally argued in favour of group or area safety representative. There is, formally, no limit to the number of safety representatives a union can appoint to cover a workplace.

Although individual workers now have a legal right under the 1993 Trade Union Reform and Employment Rights Act to refuse to do work which they consider dangerous, safety representatives do not have the right to stop such work (except inasmuch as they are asked to do the work themselves). Thus, a Union of Construction, Allied Trades and Technicians (UCATT) safety representative was able to take a successful claim to an industrial tribunal when he was sacked after calling in HSE inspectors to investigate fumes on the site where he was working.

UK and European law

The possibility originally offered by the 1974 Health and Safety at Work Act, allowing for non-union safety representatives, and the post-Cullen situation in the offshore oil industry did not undermine support for the 1977 Regulations. In part this reflects the satisfaction of both employers and unions with a system of safety representation which reflected existing industrial relations structures. However, changes in the labour market have combined with European law to suggest that the 1977 Regulations are no longer sufficient to satisfy the need for worker representation on safety at work.

The Framework Directive, implemented in Britain through the Management of Health and Safety at Work (MHSW) Regulations, provides rights to *all* workers, in accordance with national laws and practices. Some of those rights were incorporated in the MHSW Regulations, but only insofar as they were provided through safety representatives under the 1977 Regulations. The rights provided by the Framework Directive are granted to individual workers and their representatives. The Directive and recent decisions of the European Court (principally, upholding complaints by the Commission against the UK Government for failing properly to implement the Collective Dismissals 1975 and the Acquired Rights 1977 Directives) rely on such individual rights being provided through workers' representatives.

On 1 October 1996 new rights came into force under the Health and Safety (Consultation with Employees) Regulations, which provided all workers with rights to information, consultation and representation over health and safety. The TUC welcomed this extension of new rights, which will cover an estimated 10 million people, but believes that they will not adequately impact on injury rates.

The new rights cover workplaces without union recognition, recognized workplaces will continue to be covered by the 1977 Regulations. The new rights give the employers the choice of whether to consult workers directly or through representatives over health and safety. In union-recognized workplaces, the 1977 Regulations allow the workforce, through their union/s, to decide how they want to be consulted.

Any such representatives elected by the workforce under the new regulations do not have the power that union safety representatives have to inspect the workplace and investigate accidents. Nor do they have the power to require the creation of a workplace safety committee. Workers without a recognized union will therefore be getting second-class safety rights—something the TUC opposed while the new regulations were being developed, and an issue on which it is currently taking legal advice.

The reason for this two-tier system is that the Conservative Government in power up to 1997 did not want to allow workers to have the right to determine their own consultation arrangements. That would be too much like allowing workers the right to be represented by a recognized trade union, and the government resisted that too. We are concerned also that some employers believe that the new regulations allow them to ignore existing union involvement in safety representation, which they do not.

The TUC is also concerned at the implications for safety rights with the development over recent years of new forms of self-employment, particularly in the construction industry (as well as, for example, agriculture and broadcasting), where working people have self-employed status for tax purposes (largely to the benefit of their employer) but are, to all extent and purposes, still employees in that they are subject to detailed instruction in the performance of their duties.

We have noted, with satisfaction, the Court of Appeal judgment in February 1995, which held that a self-employed builder was, in terms of health and safety law, an employee of the people who told him what to do, and the new regulations on employee consultation make it clear that such workers are covered. However, we are not so sure that this view is shared by the generality of employers. HSE inspectors need to make this clear when they inspect construction sites. It also needs to be made clear to employers and principal contractors that contractors, who can work for many years alongside and under exactly the same conditions as permanent employees, also come within the scope of the regulations. The Construction (Design and Management) Regulations 1994 (CDM) do not affect the Consultation of Employee Regulations, merely the way in which they may be brought into effect.

Roving safety representatives

The 1977 Regulations, and the 1996 Regulations, embody the principle that workers should be represented by one of their own—an employee at the relevant workplace or in the same organization. In general, the TUC agrees that this is the best way to provide for worker representation, ensuring that the safety representative is practicably accountable to, and experienced in the work of, the people they represent. However, there are certain groups of workers for whom this form of representation is not necessarily appropriate. The 1977 Regulations provide for workers in theatres, for example, to be represented by safety representatives appointed by the relevant union but who are not employees at the theatre concerned, because of the peripatetic and often temporary nature of employment in that field. Commonly, such safety representatives are paid officials of the union or experienced lay officers.

The TUC believes that, in introducing Regulations to cover all workers, there is a need to extend the provision for such 'roving' safety representatives to cover other sectors where similar conditions apply. We are particularly concerned about small construction sites and the agricultural sector, where the Transport and General Workers Union (TGWU) established a pilot project of roving safety representatives on a voluntary basis with European Union funding and HSE support.* The TUC is also examining the scope for 'cyberspace' safety representatives (again, paid union officials or senior lay officers) to provide on-line safety representation for teleworkers through the Internet.

As they were not included in the 1996 Regulations, the TUC will continue to press the Health and Safety Commission to consider the possibility of extending safety representation rights to include such roving safety representatives, especially where workforces are fragmented. They should also be able to support lay safety representatives in the smallest workplaces where full safety representation might be impractical.

However, the TUC does not believe that the construction industry has to wait until the introduction of legislation. CONIAC should be taking positive steps to encourage the establishment of a system of roving safety representatives. The support of CONIAC would enable positive discussions between both sides of industry on this issue.

What difference does a union make?

Reilly *et al.* (1995) used 1990 Workplace Industrial Relations Survey (WIRS) data to examine the relationship between workplace injuries and the involvement of union safety representatives in UK manufacturing

*The TGWU initiative, launched in December 1995, was conducted in the South East and Southern regions, with safety representatives (lay members of the TGWU's agricultural section) acting as contact points for members with safety problems or questions, backed up by a union helpline and contacts with the local HSE. Where farmers agreed, these representatives made farm visits. This is reported on by Walters (1997).

establishments. The study found that 'joint consultative committees, with all employee representatives appointed by unions, significantly reduce workplace injuries relative to those establishments where the management alone determine health and safety arrangements'.

The study examined the 1990 Workplace Industrial Relations Survey 3 (WIRS3) data to identify the rate of a certain number of listed industrial injuries per thousand employees over the previous 12 months, and then compared these data with the method by which safety was organized at each establishment (varying from safety committees where all employee representatives were union nominees—essentially the 1977 Regulations system—through safety committees made up of non-union nominees, through to all decisions being taken by management). The injuries concerned were essentially serious incidents requiring the victims to be off work. The data show that the injury rate was highest where no workforce representation or consultation existed: 10·9 injuries per thousand. Non-union representation of workers, but with a joint management–employee committee, produced an injury rate of 6·1 per thousand. But the best result came from the standard 1977 Regulations model, with 5·3 injuries per thousand. These figures suggest that full union involvement in health and safety cuts the number of injuries by more than half.

Given the HSE's analysis of the costs of work-related injury and ill-health, this finding could be extrapolated to suggest that the costs of poor workplace safety could be cut by more than a half by the spread of union-based safety committees, and that would suggest a reduction from up to £16 billion a year to less than £8 billion. Clearly, this sort of extrapolation (which assumes that workplace ill-health will be as responsive to safety representation as injuries) could be dangerous, but the clear indication from this one piece of research suggests that the financial benefits of proper health and safety rights to representation could be substantial. Indeed, on these figures, even non-union safety representation would cut workplace injury rates, albeit by less than the union model.

Apart from Walters and Gourlay's 1990 report and the recent analysis of WIRS data by Reilly *et al.* (1995), there is little published analysis about the current coverage and effectiveness of union safety representatives. The TUC believes that the importance of workplace safety representatives is such that there ought to be far more research into their numbers, their activities and their effectiveness. The interesting conclusions drawn from WIRS3 suggest that, as a first step, both the Labour Force Survey and the Workplace Industrial Relations Survey could be used to produce far more detailed information on safety representatives. The research suggests that trade union organization and the integration of safety representatives into union structures are key elements of effective safety representation, as is a committed management. If the Regulations were undermined by the removal of union recognition and therefore the unions' rights to appoint safety representatives, this would do nothing to improve the position of workers. It would weaken the position of workers where there was no recognized union. But it would also weaken the position where unions are recognized but not strong.

The concerns of construction safety representatives

The TUC, over the summer/autumn of 1996, undertook a major survey of all trade union safety representatives. This survey has given the TUC a true picture of the concerns facing trade union members in the workplace.

Seventy per cent of construction safety representatives who replied to the survey said that slips, trips and falls were still the main hazards of concern to their workmates, followed closely by noise (67%), back strain (60%) and dust (59%). The survey showed that little progress had been made in eradicating the age-old hazards of construction work. The safety representatives blamed a mixture of job insecurity and lack of safety inspections for the persistence of traditional hazards in the industry. One third of the construction safety representatives who responded to the survey reported that they had never seen an inspector on their site. Other issues raised by the safety representatives in construction included solvents (50%), stress or overwork (45%), asbestos (23%) and dermatitis (23%).

Proposals for action

After two decades, the safety representatives system is in need of development. The TUC believes that, despite a generally hostile climate, safety representatives have proved their importance—but they still have a great deal of unfulfilled potential.

The main areas where change is needed are in:

- strengthening the support offered to safety representatives by trade unions and others
- improving the information available on safety representatives, initially through the Labour Force Survey and the Workplace Industrial Relations Survey
- adapting to changing health and safety requirements, especially dealing with risk assessments, and involvement generally in the management of an organization's safety culture.

For its part, as revealed earlier, the TUC has already undertaken a major survey of union safety representatives' views and activities (likely to be an annual event). Training for safety representatives will continue to be provided, despite the withdrawal of the government grant, to over 14 000 new safety representatives every year (along with short refresher/updating courses for experienced safety representatives). Finally, safety representatives' need for up-to-date information will be met in the short term with a new edition of the TUC handbook on health and safety, *Hazards at Work*, and ultimately through a strategy designed to provide safety representatives with access to information through the Internet.

The TUC will be holding further discussions with the construction industry trade unions to establish how these initiatives can be developed for their particular interests.

What of the future?

The TUC believes that there is a need to develop the undoubted role of safety representatives in the health and safety management system of the construction industry.

A research project is needed which would:

- ☐ identify the cost of accidents and ill health in the construction industry, including the cost to the national economy (this could be done using the LFS and WIRS)
- ☐ identify the extent and the role that trade union safety representatives play in the health and safety system of the industry
- ☐ establish the role safety representatives have had in improving the health and safety performance of the construction industry and identify examples of good practice.

The CONIAC Strategic Plan recognized the role that the HSE inspectorate plays in ensuring the development of good health and safety practice and the implementation of the CDM. Within this context it needs to be established what use the HSE inspectorate makes of safety representatives when visiting sites, what ways in which they should maintain contact with construction trade union safety representatives and what steps could be taken to improve that relationship.

The implementation of the Consultation Regulations will have serious implications for the construction industry and this should be especially reflected in the work of the HSC as part of its ongoing commitment to implementing the CDM. A major element of this will be the training of elected representatives. But also important will be the inspectorate's response to enforcing this legislation. There needs to be monitoring of the advice and enforcement activity undertaken by the inspectorate and evaluation of its effect.

The HSC has in the past recognized the important role that trade union safety representatives should be playing in the construction industry. The Commission would be the eminent body to establish a forum which would bring together employers and trade unions to discuss the development of the roving safety representative. All this, though, will take substantial resources to ensure that the 1977 Safety Representatives and Safety Committees Regulations and the 1996 Consultation Regulations are fully implemented in the construction industry. The TUC will continue to press government, employers and the construction industry generally to invest in the improvement of construction safety by supporting trade union safety representatives.

Bibliography

Davies, N.V. and Teasdale, P. (1994). *The Costs to the British Economy of Work Accidents and Work-related Ill Health.* HSE, Sudbury.

DTI, ACAS, Economic and Social Research Council (1990). *Workplace Industrial Relations Survey.* Policies Studies Institute, Grantham.

Health and Safety Executive (1994). *Labour Force Survey 1993/4.* HSE, Sudbury.

Health and Safety Executive (1996). *Labour Force Survey 1995/6.* HSE, Sudbury.

New Statesman and Society (18 November 1994). Research by Colin Whitston and Jeremy Waddington.

Reilly, B., Paci, P. and Holl, P. (1995). Unions, safety committees and workplace injuries. *British Journal of Industrial Relations*, **33**, No. 2, 275–288.

Walters, D. (1997). *The Role of Regional Health and Safety Representatives in Agriculture: An Evaluation of a Trade Union Initiative on Roving Safety Representatives in Agriculture.* HSE, Sudbury, HSE Contract Research Report 157/1997.

Walters, D. and Gourlay, S. (1990). *Statutory Employee Involvement in Health and Safety at the Workplace: A Report of the Implementation of the Safety Representatives and Safety Committee Regulations 1997.* HSE, Sudbury.

13. Growing a safety culture

John Anderson, *Consulting Civil Engineer, Chester*

Introduction

Managing safety in any industry is a tricky business at the best of times—'safety' was defined by one speaker at a recent conference as 'a complicated way of achieving a "non-event" outcome'. Clearly nobody wants accidents, incidents or losses of any kind, but we all have to live with a degree of risk in our lives (whether at work or elsewhere), and there will never be unlimited resources to eliminate or reduce every potential accident or loss situation. Doing the best with limited resources involves targeting the risks that matter and seeking out new and better ways of risk elimination and risk control. The most important part of these resources is the skills and abilities of the persons involved.

The very essence of the construction industry is the achievement or completion of a 'task' or a 'project'. The client or owner is paying for something he wants built, and he buys in the expertise of organizations and individuals to achieve the desired end product. So if there is any obvious 'culture' in the construction industry, it is a 'task culture' with a distinct focus on the achievement of end products. This in turn breeds 'doers'—be they designers or constructors. Effective health and safety management has no option but to fit in with this pervasive 'task culture' and the individuals and organizations that are necessarily focused on production.

The concept of organizations and/or individuals having a quantifiable 'safety culture' is relatively new. There are, as yet, no definitive ways of measuring 'safety culture' nor any obvious benchmarks to aim for, so this chapter cannot contain easy answers or 'checklists' for those seeking 'instant solutions' to improve poor health and safety performance. Safety culture is, however, a potentially valuable and important concept which is attracting a good deal of research interest—particularly within the energy industries. The construction industry needs to keep an eye on developments and be prepared to take ideas from the emerging research findings that might be of future benefit.

This chapter will:

- ☐ discuss the factors that have led to the present pattern of accident trends in construction, and how the record might proceed
- ☐ consider the subject of 'culture' on a number of different levels
- ☐ define safety culture and review the research record to date and consider what might be of interest to the construction industry

- □ look to how the construction industry might take this matter forward, bearing in mind the number of 'barriers' or 'obstacles' that might be in the way of progress.

Long-term broad stages in accident reduction in construction

Over the long term there has been a considerable improvement in the reduction of fatal accidents in construction, and also in the recorded accident incidence rates per numbers employed. This is surely to be expected as knowledge increases and information and guidance become more available. It seems possible that over the very long term there may have been several periods when the accident record has 'plateaued', until something or someone has come along and encouraged the trend toward a new, lower level. Figure 1 postulates that there might well have been three such plateaux in the past, and that there is a potential at this moment in time for a further 'breakthrough'.

The first plateau may have been reached prior to the industrial revolution. The early accident patterns are likely to have reflected how common construction materials like stone, masonry and timber were traditionally used and assembled. Constructors and designers would rely to a great extent on 'rules of thumb' and methods of trial and error, and a common experience would be built up of what did and did not work as

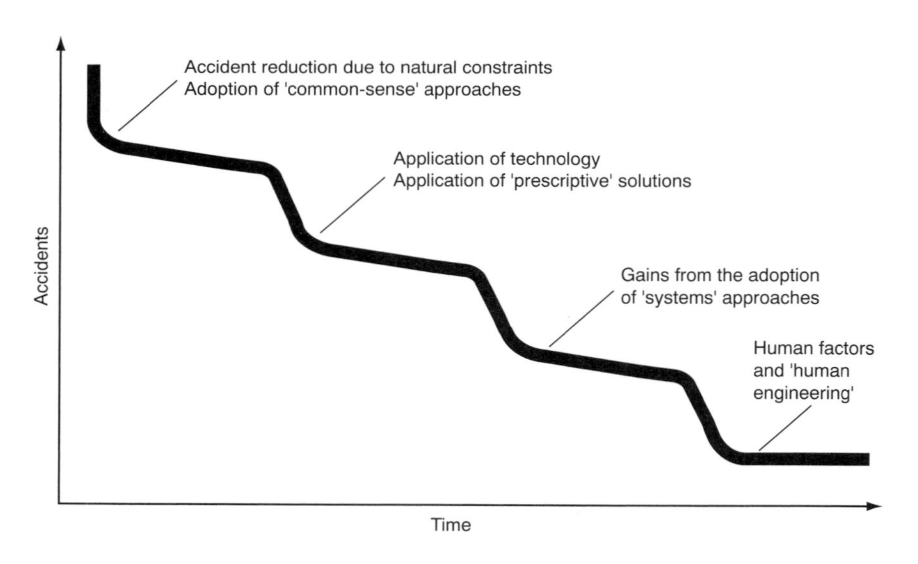

Fig. 1. 'Stages' of accident reduction over time

opposed to what was necessarily safe or not. For example, in tunnelling the common material for a long time for blasting of rock was gunpowder. The first method of use was 'plug shooting' (Sandstrom, 1963), and it is recorded that many miners were killed by the plug. Other methods were developed as experience and knowledge were shared; by 1862 Alfred Nobel had discovered new, more reliable and safer methods, and in particular the means of exploding nitroglycerine, after, it is said, 'blowing a succession of laboratories to bits'.

Following the advent of the Victorian era, and the more rigorous application of the laws of physics to the theory and practice of civil engineering, more codes and standards and guides to sound practice became available. No doubt the formation of the Institution of Civil Engineers with its learned society activities played its part. This was also the time of the framing of legislation concerned with conditions at work. Most of this legislation was of a prescriptive nature and concerned with specific places of work and specific hazardous work activities. This pattern of legislation followed the identification of accident patterns and the need to frame rules or regulations about obvious shortcomings of the physical working environment. If accidents were to be attributed to 'unsafe acts' as opposed to 'unsafe conditions', then the remedy for 'unsafe acts' was probably thought to be instruction and training. It was not until 1937 that specific regulations with regard to the dangers of construction work were enacted, and even then the emphasis was on safety as opposed to health. But even with substantial strides in the identification of accident types, advances in the application of technology and the formulation of prescriptive solutions to address unsafe conditions, accident reduction could only go so far.

The next advance was the work of the Robens Committee in 1972, which led to the drawing up of the Health and Safety at Work etc. Act (HSWA) 1974. Section 2(a) required: 'the provision and maintenance of systems of work that are, so far as reasonably practicable, safe and without risks to health'. This, rightly, caused a substantial amount of rethinking to be done as it was not just the individual parts that mattered in themselves, but the whole way the construction *system* was put together. A *systems approach* (which could be described as a set of related elements designed to achieve some specific purpose or goal) can achieve much more if there is a broad enough perspective. Thinking and controlling one part or several individual parts of one process was not enough. The construction industry has always had to adopt a systems approach by the very nature of our work. We have to think about the related activities of planning, designing, constructing and maintaining. Two more recent legislative developments have reinforced the systems approach—the advent of Regulation 4 of the Management of Health and Safety at Work Regulations 1992 and the Construction (Design and Management) (CDM) Regulations 1994.

Regulation 4 of the Management Regulations spells out in more detail what are to be the minimum parts of the system: 'Every employer shall make and give effect to such arrangements as are appropriate, having regard to the nature of his activities, and the size of his undertaking, for the effective planning, organisation, control, monitoring and review of the preventative measures' (i.e. risk elimination and control). This Regulation requires employers to have a system in place to cover health and safety issues which contains, as a minimum, the management system elements of planning, organizing, control, monitoring and review. The legislation is now aimed at prescribing the *processes* that employers should follow. Prior to 1974 duties rested largely with 'employers of workmen' engaged on defined 'building operations and works of engineering construction', and while the HSWA extended the duties to other employers (such as consulting engineers who had employees at risk on construction sites), no specific duties rested on those who commissioned construction work or who designed the works in response to a commission. That was changed by the introduction of the CDM Regulations. This particular system of legal duty holders—clients, designers, specifiers, constructors, manufacturers and suppliers of construction plant, materials and those maintenance and other activities on the final product was now complete. Working in systems has produced, and will continue to produce, benefits for many years to come.

But as further safety gains might diminish, is there a new frontier that can be crossed? James Reason (1990) states that 'Modern technology has now reached a point where improved safety can only be achieved through a better understanding of human error mechanisms'. One might well argue that this could be true for certain sectors of advanced manufacturing industry, and perhaps for the nuclear power energy sector. In construction there still appears to be ample opportunities to improve health and safety performance, but the idea of looking for new gains in terms of examining the questions of human behaviour and in particular the likelihood of human error is well made. Wrong decision making by the human part of the any safety-critical system is obviously important. If we can get the technology right; the systems right; and avoid error or misjudgement by human participants—surely that will that be as near-perfect a system as we can possibly manage?

Possible components of 'culture'

If 'culture' is about anything, it is about both individuals and groups of people who, for whatever reason, share values, attitudes and beliefs. These words, 'values, attitudes and beliefs', are commonly used in literature on 'culture' but on their own they miss out on the dimension of behaviour or action. A group of people in construction subscribing to high safety values etc. may be under pressure to compromise these 'ideals' given immediate and severe time or resource constraints on site. Perhaps accidents have their roots in the extent to which a safety culture exists (or does not exist, as the case may be).

If one was to compose a list of the possible components of a safety culture, such a list might contain the following:

- values (e.g. the conviction that certain ways of achieving goals or performing the tasks to achieve these goals is right and proper)
- attitudes (e.g. towards safety, and its importance in the scheme of things)
- norms (e.g. set procedures, rules or established customs and practices, and the extent to which they embrace safety)
- moral and ethical considerations (which could be paramount in certain professions)
- beliefs (how things are or should be)
- perceptions (how things might work out—an important consideration in safety matters)
- professional considerations (e.g. the extent to which professional duties and standards are imposed on members of professional bodies—these bodies being legitimized by statute or royal charter, and enjoying public support).

Before attempting to put these constituents into some sort of meaningful definition, it is clear that different cultural dimensions can exist on a number of levels. Research (Hofstede, 1980) into the subject has looked at the international differences with some surprising results. The researchers described three indices:

(a) *Power Distance Index*, which measured the extent to which members of society accept that power is distributed unequally.
(b) *Uncertainty Avoidance Index*, which measured the degree to which people feel threatened by ambiguous situations, and create beliefs and institutions that try to avoid uncertainty.
(c) *Individualism Index*, which measures the extent to which people believe that their main concern in life is the well-being of themselves and their immediate families (as opposed to an orientation towards a wider grouping with more extended responsibilities).

Both (b) and (c) can be important in the more detailed consideration of safety culture. To illustrate the variability of findings under these three indices, see Table 1.

Societies in countries with a high Power Distance Index score (i.e. with a strong sense of injustice about the inequality of power distribution) may pose problems for achieving consensus and compromise, whereas those with low scores may not easily accept a strong leader with an autocratic

Table 1. A selection from the published research results

Country	Power Distance Index	Uncertainty Avoidance Index	Individualism Index
Austria	11	70	55
Denmark	18	23	74
France	68	86	71
Germany	35	65	67
Great Britain	35	35	89
Greece	60	112	35
Japan	54	92	46
Spain	57	86	51
USA	40	46	91
For the complete 40 countries surveyed			
Mean	52	64	50
SD	20	24	25

Source: Hofstede (1980, p. 315)

style. Even more interesting are the Uncertainty Avoidance Index scores, where higher than average results (e.g. Greece and Japan) would point to a national tendency to produce rules and regulations to fulfil tasks whereas low-scoring countries (including Great Britain) feel reasonably comfortable with a certain level of uncertainty. In addition, these latter countries may well resist the idea of excessive rules and bureaucracy (including that imposed for health and safety reasons). A country high on the Individualism Index (e.g. USA) will be keen to 'get on with the practicalities of the job' and show less concern and interest in peripheral activities. Comparisons of health and safety legislation and preferred approaches used within certain countries tend to support the views expressed by the researchers.

Defining 'safety culture'

It is generally accepted that the term 'safety culture' was first researched for the nuclear industry, and it received considerable attention following the investigation into the Chernobyl Power Station incident. The most important document on safety culture at the present time is *Third Report; Organising for Safety* by the Advisory Committee on the Safety of Nuclear Installations (ACSNI)—Human Factor Study Group (Health and Safety Commission, 1993). This publication (paragraph 78) suggested the following working definition: 'The safety culture of an organisation is the product of the individual and group values, attitudes, perceptions, competencies, and patterns of behaviour that determine the commitment to, and the style and proficiency of, and organisation's health and safety management'. The authors also add: 'Organisations with a positive safety culture are characterised by communications founded on mutual trust, by shared perceptions of the importance of safety, and by confidence in the efficacy of preventative measures'.

In a recent paper, Lee (1997) summarized some of the research that has taken place in the nuclear industry, where it was found that the characteristics of low-accident nuclear plants include:

- ☐ a senior management that is strongly committed to safety; giving it a high priority; devoting resources to it and actively promoting it personally
- ☐ a strong focus on safety by the organization and all its members
- ☐ a high level of communication between and within levels of the organization where safety matters are discussed and exchanges are less formal as well as more frequent; managers do more walkabouts and are more 'visible'
- ☐ a management leadership style that is democratic, co-operative, participative and humanistic as distinct from autocratic and adversarial
- ☐ more and better quality training, not only specifically on safety, but also with safety aspects emphasized during skills training
- ☐ high job satisfaction, with favourable perceptions of the fairness of promotion, lay-off and employee benefits as well as task satisfaction
- ☐ good organizational learning, where organizations are tuned to identify and respond to structural change.

Scott Geller (1994) goes further in his article 'Ten principles for achieving a total safety culture', in the North American manufacturing industry context. He draws particular attention to the integrated nature of the person (knowledge, skills, abilities, intelligence, motives and personality); the environment (equipment, tools, housekeeping, hot/cold); and behaviour (complying, coaching, recognizing, communicating, demonstrating and 'actively caring').

Although the ACSNI document is written with the nuclear industry in mind, most of the findings provide practical signposts as to how the present safety culture of the construction industry might be improved. However the ACSNI researchers qualified their findings in one important respect (paragraphs 113/114):

> The most familiar point to emerge was the threatening trade off between production and safety. The latter, it is thought, requires slower and more careful work with scrupulous following of the rules. The former can be increased by faster turn around and by 'cutting corners' ... A positive

safety culture is one where safety not only wins out if there is a conflict, but where everything is done to remove the conflict ... The conditions that make for safe operation are often those that make for a good organisational climate and hence good output.

These particular comments appear to pose direct questions to certain sectors of the construction industry and how they might aspire to a positive safety culture. If the 'task focus' or emphasis on production is the first priority, does this mean that a positive safety culture is unlikely if not impossible to achieve in the construction industry's environment?

Possible barriers to a strong safety culture in the construction industry

If change and improvement are to be made from the present position, it does not seem unreasonable to attempt to describe and assess what general characteristics of the construction industry may stand in the way of a more positive safety culture. Such a list of 'barriers' might include:

- Apathy towards health and safety issues and accidents.
- The quality of some site or project management, and their ability to give health and safety the priority, resources and personal commitment it deserves.
- The complacency of top management in some companies concerning the need to improve safety performance and to perceive it as a fundamental business aim.
- The fact that in much of the industry reactive accident prevention is looked upon as easier and preferable to proactive risk reduction.
- The inadequate manner in which many accidents and incidents are investigated, and the fact that lessons are poorly 'learned' by the industry as a whole. In particular cases there is often a lack of root cause analysis, and a failure to look at behavioural issues and the environment under which 'human errors' were said to have been made. Lucas (1990) put it this way: 'Wise men learn from others' harms; fools by their own'.
- It is sometimes the case that safety officers and safety committees, while doing good work, effectively distance the subject from the key responsibilities of management.
- The industry does not give enough forward thought to where and when in safety-critical construction operations human errors can be made; what consequences might result; and what preventative interventions or strategies might be most effective.
- The lack of a safety culture within the technical educational system which produces the engineering managers of the future. The

understanding of how risks can be identified and managed via elimination and control should be an integral part of the curriculum of technical degrees.

- The fact that in some sectors a 'macho' culture is still allowed to prevail. It seems impossible to conceive that these sorts of risk takers are unaware of the many real risks to life and limb in construction. It must be that they accept these risks as a trade-off for employment or reward, and that, as far as they are concerned, production is the first goal.
- The practice in parts of the industry that the reward systems are based exclusively based on production targets or piece work. (i.e. there is a possible conflict between safety and productivity). There is a need to restructure some work situations so that rewards are not given for taking risks but more for following necessary safety precautions.

Given the size and diversity of the construction industry, these barriers to progress may appear daunting, yet it is well known that high health and safety performance can be achieved as the following case studies illustrate.

Case A

This concerned a substantial building project north of London for a large multinational client. The client made it absolutely clear that it demanded, and was willing to pay for, the very highest standards of construction site safety and that such systems to achieve that end were to be put in place. One might say that a particular safety culture was 'imposed' by the client, but put in place by the contractors involved who adapted their organizational structures and working practices to meet the standards demanded. The client sponsored an extensive site-wide safety incentive reward system which was seen by some outsiders to be particularly generous. Gibb and Forster (1996) reviewed the safety motivation on this particular project and concluded that the incentive scheme was used to good effect, but that it had to be part of a project-wide safety culture.

Case B

In this case—a tunnelling contract in the London area—the high health and safety performance was the direct result of a particularly enthusiastic and single-minded site agent of the contractor. By the very force of his personality and his knowledge of accident situations and how they could be avoided, his particular project on that site was run with safety as the first priority. He was backed up by his senior management, who were fully aware of the financial and productivity benefits of effective risk elimination and control, and by a particularly effective safety adviser who did just that—providing first-class advice and guidance when and where it was needed. On this site there was no conflict or division between production and safety—the two were considered as one fully integrated activity.

Case C

A small steeplejack company in Scotland had a modest 'headquarters' office in a rural village but was renowned for their work all over the country on church steeples, cooling towers, chimneys, etc. On most occasions they were called in as a result of some 'problem', such as on the discovery of damage or structural deterioration of the structure. The degree of uncertainty about what they might find and how they might proceed was often considerable. Many persons may perceive their work on ladders high on exposed towers and steeples as particularly 'risky', but the opposite was the case with this company. The whole structure of the company, small though it was, was geared to a very high safety culture. Every aspect was included—staff recruitment, plant, equipment, education, training, systems of work, etc.—even down to the detailed selection of footwear. In this case their lives depended on what they did; how they did it; and how they helped and cared for each other in their unique hazardous work environments.

Pointers to the future in construction

The 1993 ACSNI publication does provide some advice about action that could be taken specifically to review and promote safety culture (see pp. 49–56 of their document). The first piece of advice is that a 'step-by-step approach is essential', and that the major steps of any such plan should be:

- review the existing safety culture
- decide the aspects that have the highest priority for change
- decide on actions that may change those aspects, and to launch these actions
- repeat the previous three steps indefinitely.

The results of new research have added some detail to this framework. Lee (1997) favours attitude surveys of staff to assess the safety cultures and to help design therapeutic measures. The process, as applied to Sellafield in 1992, had four stages.

Stage 1 Approximately 50 people were brought together comprising a representative cross-section of the workforce into five 'focus groups'. The purpose of these groups was to identify those safety aspects of working which were of concern to them, and to bring to the surface a range of beliefs, attitudes and feelings that might otherwise be inaccessible.

Stage 2 The records of the focus group discussions were used to help compile and structure an extensive draft questionnaire which was piloted with a systematically selected proportion of the whole workforce.

Stage 3 The pilot questionnaire was refined and then issued to the whole workforce. In the 1992 Sellafield exercise the author reports that 5295 completed questionnaires were returned, representing an 85% response rate.

Stage 4 The results were subjected to detailed analysis to extract the underlying factors which Lee (1997) grouped under nine 'domains' and 19 associated 'factors/attitudes' as in Table 2.

In a parallel paper, Rycroft (1997), of British Nuclear Fuels Ltd at Sellafield, argued that any technique used for judging or measuring safety culture must be able to identify:

- ☐ Perceptions. What are the workforce's opinions on the safety of their workplace? Do they feel safe?

Table 2. The 'domains' and 'factors/attitudes' of Lee (1997)

Nine domains	Nineteen factors/attitudes
Safety procedures	confidence in the safety procedures
Safety rules	personal understanding of the safety rules perceived clarity of safety rules
Permit to work systems (PTW)	confidence in the effectiveness of PTW general support for PTW perceived need for PTW
Risks	personal caution over risks perceived level of risk at work perceived control of risks as a whole
Job satisfaction	personal interest in the job contentment with the job satisfaction with work relationships satisfaction with rewards for good work
Participation/ownership	self-participation in safety procedures perceived source of safety suggestions perceived source of safety actions perceived personal control over safety
Design	satisfaction with the design of the works
Training	satisfaction with training
Selection	satisfaction with staff suitability

- ☐ Attitudes. How do the workforce see their and others' responsibility for safety? Are they proactive or passive? Is safety at work somebody else's job?
- ☐ Strengths and weaknesses. What do they do well or badly? How do they compare with companies with high health and safety performance?
- ☐ Beliefs versus behaviours. Is what they say what they do?

Mearns *et al.* (1997), in reporting on the results of safety culture research in the offshore oil industry for the Health and Safety Executive, has similar headings to the above, but a further questionnaire was devised for contractors employed by the operating companies. This is obviously one step nearer the multicontractor system in construction. In this case there was an identification of (a) their own management's commitment to safety, (b) their own company commitment to safety and (c) the confidence they had in the operating company with regard to safety. Thompson (1997), in a paper which has strong resonances for the construction industry, urges that the drive to higher levels of safety culture and lower accidents and incidents is by moving through three stages of safety culture: dependent, independent and interdependent (see Fig. 2). In the interdependent stage it is the team role that is the dominant factor, where everyone becomes a team contributor working towards team goals—taking responsibility and not ignoring the safety of someone else.

Conclusion

The proponents of a strong safety culture will argue that:

- ☐ it contributes to business competitiveness (good health and safety performance is indeed good business)
- ☐ there are gains to be had in productivity and quality
- ☐ it will result in improvements in industrial relations
- ☐ there are likely to be gains in job and task satisfaction which in themselves will help to retain and attract good staff and add to the 'standing' of the organization
- ☐ it will encourage a wider 'ownership' of health and safety as part of all day-to-day business
- ☐ it will lead to a reduction in accidents and incidents via improved risk management.

It is clearly a 'good thing'. Quite how it should be researched, evaluated and improved within the construction industry is, as yet, unclear, but the gains that have been made elsewhere just cannot be ignored. Robens (1972) was right to remind us, more than 25 years ago, that 'the most important single reason for accidents is apathy'.

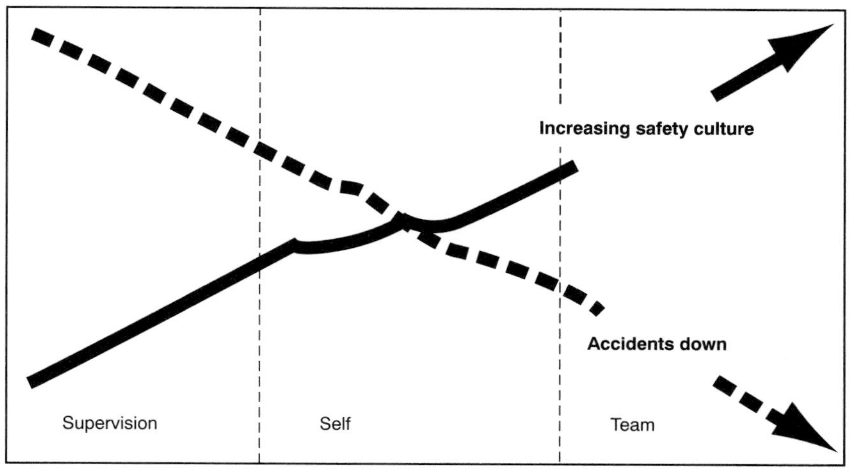

DEPENDENT	INDEPENDENT	INTERDEPENDENT
Supervision emphasis	Personal commitment	Team commitment
Fear/discipline	Personal standards	Team contributor
Laws/rules	Self-managing	Team goals and objectives
Set procedures	Care for self	Care for others

Fig. 2. The three stages of safety culture after Thompson (1997)

References

Committee on Safety and Health at Work (Robens Committee) (1972). *Safety and Health at Work*. HMSO, London.

Gibb, A.G.F. and Forster, M. (1996). Safety motivation; evaluation of incentive schemes. *Proceedings of the First International Conference of CIB Working Commission, Portugal, September 1996*. In *Implementation of Safety and Health on Construction Sites* (eds L.M. Alves Dias and R.J. Coble). Balkema, Rotterdam, pp. 405–416.

Health and Safety Commission (Advisory Committee on the Safety of Nuclear Installations—Human Factors Study Group) (1993). *Third Report; Organising for Safety*. HSE, Sudbury.

Hofstede, G. (1980). *Cultures Consequences*. Sage, London.

Lee, T. (1997). How can we monitor the safety culture and improve it where necessary? *Proceedings of a Conference on 'Safety Culture in the Energy Industries'*, Aberdeen, organized by Energy Logistics International Ltd, Cookham.

Lucas, D. (1990). Wise men learn by others harms, fools by their own; Organisational barriers from learning the lessons from major disasters. In *Safety and Reliability in the 90s: Will Past Experience or Prediction Meet our Needs?* (eds M.H. Walter and R.F. Cox). Elsevier, London.

Mearns, K. *et al.* (1997). Measuring safety climate on offshore installations. *Proceedings of a Conference on 'Safety Culture in the Energy Industries', Aberdeen*, organized by Energy Logistics International Ltd, Cookham.

Reason, J. (1990). *Human Error*. Cambridge University Press, Cambridge.

Rycroft, H.S. (1997). Developing a safety culture in a changing organisation. *Proceedings of a Conference on 'Safety Culture in the Enery Industries', Aberdeen*, organized by Energy Logistics International Ltd, Cookham.

Sandstrom, G. (1963). *The History of Tunnelling*. Barrie and Rockcliffe, London, ch. 12.

Scott Geller, E. (1994). Ten principles for achieving a total safety culture. *American Society of Safety Engineers*, Sept., 18–24.

Thompson, P. (1997). Developing a safety culture in practice-interdependency and involvement. *Proceedings of a Conference on 'Safety Culture in the Energy Industries', Aberdeen*, organized by Energy Logistics International Ltd, Cookham.